Identification and augmentation of a civil light helicopter: transforming helicopters into Personal Aerial Vehicles

Stefano Geluardi

Identification and augmentation of a civil light helicopter: transforming helicopters into Personal Aerial Vehicles

Dissertation

Engineering Ph.D. School "Leonardo Da Vinci"
Ph.D. Program in Automation, Robotics
and Bioengineering

University of Pisa
Department of Information Engineering
and
Max Planck Institute for Biological Cybernetics
Department of Human Perception, Cognition and Action

by

Stefano GELUARDI
2016

Bibliografische Information der Deutschen Nationalbibliothek

Die Deutsche Nationalbibliothek verzeichnet diese Publikation in der Deutschen Nationalbibliografie; detaillierte bibliografische Daten sind im Internet über http://dnb.d-nb.de abrufbar.

ISBN 978-3-8325-4366-2

Logos Verlag Berlin GmbH
Comeniushof, Gubener Str. 47,
10243 Berlin
Tel.: +49 (0)30 42 85 10 90
Fax: +49 (0)30 42 85 10 92
INTERNET: http://www.logos-verlag.de

Date of examination:
   July 5, 2016

PhD Course Chairman:
   Prof. dr. ir. A. Caiti

Tutors:
   Prof. dr. ir. L. Pollini
   Dr. ir. F. M. Nieuwenhuizen

Committee members:
   Prof. dr. ir. A. Landi
   Prof. dr. ir. C. Colombo
   Prof. dr. H. H. Bülthoff

"Once you have tasted flight,
you will forever walk the earth
with your eyes turned skyward,
for there you have been,
and there you will always
long to return"

Leonardo Da Vinci

Summary

Identification and augmentation of a civil light helicopter: transforming helicopters into Personal Aerial Vehicles

Stefano Geluardi

I N the past years congestion problems have led regulators to consider the implementation of new concepts for the general public transportation. The possibility to exploit the vertical dimension, by moving commuting traffic from the ground into the air, motivated the European Commission to fund a four year project, the myCopter project. The aim was to identify new concepts for air transportation that could be used to achieve a Personal Aerial Transport System (PATS) in the second half of the 21st century. In this new system, Personal Aerial Vehicles (PAV) would represent the basic means of transport used by the civil community. Although designing a new vehicle was not among the project's goal, it was considered important to define response types and handling qualities PAVs should have to be part of a PATS.

Today, efforts to create PAVs-like vehicles are underway in many parts of the world. Many of these attempts consist of creating aircrafts with vertical take off and landing capabilities. This feature is particularly important since it avoids the need to build infrastructures such as airports or roads. Although many of these vehicles have succeeded in combining benefits of conventional ground-based and air transportation, so far none of them has achieved mass production. The main cause lies in the financial capabilities and the piloting skills these vehicles still require.

In this thesis a different approach was considered, in which civil light helicopters were proposed as possible PAVs candidates. The advantage of using helicopters is that they are well known vehicles massively produced all over the world. Furthermore, civil light helicopters possess many properties a PAV should have (e.g. size, weight, number of seats, vertical take-off and landing capabilities). However, they are still today a niche product because of costs and long trainings necessary to obtain a pilot license. This makes this solution affected by the same issues that prevent many new vehicles from being broadly accessible to the general public.

The goal of this thesis was to face these issues by investigating if civil light helicopters can be transformed into PAVs through system identification methods and control techniques. The transformation was envisaged in terms of vehicle response and handling qualities. In other terms, it was studied whether civil light helicopters can be transformed into vehicles anyone can fly with little training.

An initial motivation for this thesis was strictly linked to the state of the art of civil helicopters and the lack of research in the civil helicopter field. In the last decades a considerable literature has been devoted to helicopter system identification for military purposes. However, the civil helicopter field has shown very little interest in this kind of studies. Therefore, the first step was to investigate how the civil helicopter community could benefit from using system identification.

Another motivation was related to the focus of the helicopter control literature, oriented towards handling qualities improvements for professional and highly trained pilots. Understanding whether civil helicopters can be augmented to achieve PAVs handling qualities was seen as a radical change in the helicopter research field towards the direction of civil inexperienced pilots.

Three objectives were formulated to achieve the goal of the thesis. In the first objective, a civil light helicopter model was implemented for the hover condition. The second objective consisted of augmenting the identified helicopter model to achieve response types and handling qualities defined for PAVs. The third objective considered the assessment and validation of the implemented augmented systems in piloted closed-loop control tasks performed by inexperienced pilots.

To achieve the first objective, a Robinson R44 Raven II helicopter was identified in hover. The hover condition was considered well suited for the goal of the thesis as it represents one of the most difficult to perform, particularly for inexperienced pilots. First a lean and practical procedure was implemented to obtain reliable measurements of control input signals and helicopter responses. Then, a frequency domain identification method was adopted. A non-parametric model was implemented, followed by a transfer function model. Finally, a state-space fully-coupled model was identified. In this step, the implemented algorithm used to identify the state-space model appeared to be sensitive to initial parametric values and bounds. Therefore, new guidelines were proposed to limit this issue. The resulting state-space model showed good predictive capabilities and was able to capture high frequency rotor-body coupling dynamics. Furthermore, the model was assessed by a helicopter pilot while performing hover and low speed control task maneuvers in the MPI CyberMotion Simulator. The contribution of the first objective was to show that identification methods are today mature enough to be adopted for studies and applications in the civil helicopter field.

To achieve the second objective, an optimal H∞ and a robust μ-synthesis techniques were designed and compared to each other. A multi-objective optimization problem was implemented to select the parameters of the weighting functions used in the control design. Stability and performance robustness achieved by the two control techniques were tested against parametric uncertainties and external disturbances. Results showed that both classic control methods were able to modify the helicopter handling qualities to resemble those defined for PAVs. In particular the H∞ control method achieved better nominal performance results, but only the μ-synthesis control approach was able to ensure robust stability. However, none of the control techniques was able to guarantee performance requirements against the parametric uncertainties of the helicopter model. The contribution of the second objective was to prove that classic control techniques can augment helicopter dynamics to resemble response types and handling qualities defined for PAVs.

The third objective consisted of validating and comparing the implemented control systems in piloted closed-loop control tasks. To achieve this objective, an experiment was conducted in the MPI CyberMotion Simulator. Subjects with no prior flight experience were invited to perform two control tasks maneuvers: the hover and the lateral reposition. Each subject was assigned to one dynamics among the identified helicopter model, the PAV reference model, the H∞ augmented system and the μ-synthesis augmented system. An actual helicopter pilot was also invited to participate in the experiment. The pilot was assigned to the helicopter dynamics condition to allow for comparisons with the other inexperienced subjects. Results were evaluated in terms of objective and subjective workload and performance.

The first finding of the experiment was that ordinary inexperienced pilots are not able to control a helicopter model after a short training of 20 minutes, which confirmed the necessity of long trainings (about 50 hours) to obtain a pilot license. However, a 20 minutes training was sufficient to allow inexperienced pilots to fly a PAV model and perform maneuvers with good performance. Furthermore, a remarkable result was achieved by the H∞ augmented control system as no significant difference was found with respect to the PAV reference model in terms of objective performance and workload. The μ-synthesis control system performed slightly worse than the PAV reference model, but only in the vertical position of the hover maneuver. Overall, both augmented control systems succeeded in resembling PAVs handling qualities and response types in piloted closed-loop control tasks. The contribution of the third objective was to validate that helicopters can be augmented to achieve performance and workload levels comparable to those defined for PAVs. In other words, it is possible to transform helicopter dynamics into PAVs ones. Therefore, the approach proposed in this thesis represents a valid alternative to the common practice of implementing new vehicles that can achieve specific requirements like those defined for PAVs.

An additional contribution of this thesis was the design of an $\mathcal{L}_1$-adaptive control to compensate the performance degradation caused by the parametric uncertainties of the identified helicopter model. To verify the validity of the $\mathcal{L}_1$-adaptive control approach, the adaptation was applied to a simple PID-based controller, used to augment the helicopter model. The adaptive control was applied to reject the effects of uncertainties and restore the nominal behavior of the augmented system. A Monte-Carlo study was performed to validate the proposed control architecture against different realizations of the uncertain parameters. Results showed that the implemented adaptive controller can significantly reduce the effects of uncertainties on the augmented helicopter performance. Based on these results, further studies might be considered to apply the $\mathcal{L}_1$-based adaptive control to the H∞ and the μ-synthesis augmented systems implemented in this thesis.

The approach developed in this thesis validated the possibility to consider civil light helicopters as PAVs candidates. System identification methods and control strategies were conveniently adapted to build and augment a civil light helicopter model. The Max Planck CyberMotion Simulator was used to validate the augmented helicopter models in piloted closed-loop control tasks and test their capability to achieve performance and workload levels defined for PAVs. Furthermore, it was shown that the $\mathcal{L}_1$-based adaptive control technique can be used to reduce performance degradation due to model parametric uncertainties.

Starting from these results, further studies might be conducted to investigate if uncertainties in the model parameters could affect performance and workload of inexperienced pilots in piloted closed-loop control tasks. This would allow for handling qualities levels to be defined as those considered for military pilots in the ADS-33E document. Another important aspect of this thesis is the focus on hover and low speed flight regime, well suited for the proposed goal. Further studies could be considered to generalize and extend results and findings of this thesis by including other flight conditions, e.g. high speed flight regime.

Nomenclature

Acronyms

CMS   CyberMotion Simulator

CoG   Center of Gravity

CRi   Cramer-Rao bound of the ith identified parameter of the state-space model

HQs   Handling Qualities

MTE   Mission Task Element

MUAD   Maximum Unnoticeable Added Dynamics

PATS   Personal Aerial Transportation System

PAVs   Personal Aerial Vehicles

PID   Proportional Integral Derivative

VTOL   Vertical Takeoff and Landing

Greek Symbols

$\bar{\nu}_0$ trim inflow ratio

$\beta_0, \beta_{1c}, \beta_{1s}$ rotor coning, longitudinal and lateral flapping angles, $[rad]$

$\delta_{lat}, \delta_{lon}, \delta_{ped}, \delta_{col}$ helicopter control inputs (lateral cyclic, longitudinal cyclic, pedals rudder, collective lever), $[deg]$

η_{C_t} integrated perturbation thrust coefficient

Γ adaptive gain

γ Lock number

ν rotor inflow velocity, $[m/s]$

Ω rotor rotation speed, $[rad/s]$

ω frequency, $[rad/s]$

ϕ, θ, ψ fuselage angular attitude (roll, pitch, yaw) earth-fixed coordinates, $[rad]$

ρ atmospheric density, $[Kg/m^3]$

ρ_i ith identified parameter of the state-space model

σ rotor solidity

τ_f rotor flap time constant, $[s]$

ξ, ω damping and natural frequency of a second order system

Latin Symbols

a lift-curve slope, $[1/rad]$

a_x, a_y, a_z accelerometer components along the body axes (longitudinal, lateral, vertical), $[m/s^2]$

$C(s)$ adaptive low-pass filter

C_0 Carpenter-Fridovich inflow constant

g gravitational acceleration, $[m/s^2]$

L, M, N external moments about the helicopter center of gravity (roll, pitch, yaw), $[N \cdot m]$

$L_{\delta_{lat}}$ example of dimensional control derivative; $L_{\delta_{lat}} \equiv \delta L / \delta \delta_{lat}$

L_p example of dimensional stability derivative; $L_p \equiv \delta L / \delta p$

m helicopter mass, $[Kg]$

p, q, r fuselage angular rates (roll rate, pitch rate, yaw rate), $[rad/s]$

R main rotor radius, $[m]$

s Laplace transform variable

u, v, w velocity components along the body axes (longitudinal, lateral, vertical), $[m/s]$

X, Y, Z external forces on the helicopter center of gravity (longitudinal, lateral, vertical), $[N]$

x_a, y_a, z_a offset of the accelerometer package relative to the center of gravity, $[m]$

Contents

Summary   ix

Nomenclature   xv

1   Introduction   1
    1.1   Helicopters as Personal Aerial Vehicles   3
    1.2   Helicopter identification   5
    1.3   Helicopter augmentation   7
    1.4   Research motivation   8
    1.5   Objectives   8
    1.6   Approach and outline of the thesis   10
    1.7   Thesis scope   12
    1.8   Publications   13

2   Identification of a civil light
   helicopter in hover   15
    2.1   Collection of flight test data   21
    2.2   Development of the measurement setup   21
        2.2.1   Required signals   21
        2.2.2   Instrumentations for the output vehicle signals   22
        2.2.3   Instrumentation for piloted control inputs . . .   22
        2.2.4   Validation of the measurement setup for piloted
               control inputs   25

2.3   Choice of flight maneuvers   27

    2.3.1   Pilot training phase   27

2.4   Flight tests   28

2.5   Frequency domain identification method   30

    2.5.1   Non-parametric identification   30

    2.5.2   Transfer function and state-space modeling . .   32

2.6   Non-parametric identification results   34

2.7   Transfer function modeling results   37

    2.7.1   Pitch response   37

    2.7.2   Heave response   40

2.8   Extended state-space identification   43

2.9   State space modeling results   47

2.10 Validation of the state-space model   54

2.11 Conclusions   57

3 Robust control methods to augment an identified helicopter model   **59**

3.1   Helicopter model   63

3.2   Robust control theory   64

    3.2.1   H∞ control   64

    3.2.2 μ-synthesis control   65

3.3   Control design   65

    3.3.1   Control System Architecture   66

    3.3.2   Weighting Function Selection   69

3.4   Results   72

    3.4.1   Stability   72

    3.4.2   Nominal Performance   73

    3.4.3   Robust Performance   76

3.5   Discussion   80

3.6   Conclusions   81

4 Evaluation and validation of augmented control systems in piloted closed-loop control tasks   **83**

4.1   Experimental method   86

    4.1.1   Apparatus   86

    4.1.2   Independent variables   87

    4.1.3   Mission Task Elements   88

		4.1.4	Dependent measures	90
		4.1.5	Participants and instructions	92
		4.1.6	Experimental procedure	92
	4.2	Hypotheses	94
	4.3	Results	95
		4.3.1	Hover MTE objective performance and workload	95
		4.3.2	Hover MTE subjective rating	98
		4.3.3	Lateral reposition MTE objective performance and workload	98
		4.3.4	Lateral reposition MTE subjective rating . . .	101
	4.4	Discussion	102
	4.5	Conclusions	104

5 $\mathcal{L}_1$-based model following control to suppress model parametric uncertainties effects — **107**

	5.1	Helicopter Model and Augmentation System	110
		5.1.1	Helicopter Model	110
		5.1.2	Dynamic Augmentation with a Baseline PID Controller	111
	5.2	Uncertainties	113
	5.3	$\mathcal{L}_1$-controller theory and implementation	118
		5.3.1	Uncertain Plant	118
		5.3.2	State Predictor	119
		5.3.3	Adaptation Law	119
		5.3.4	Control Law	120
	5.4	Monte-Carlo Simulation and Results	122
	5.5	Conclusions	124

6 Conclusions and Recommendations — **127**

	6.1	Definition of new guidelines for state-space model identification	130
	6.2	Weighting function selection in optimal control design	131
	6.3	Assessment of augmented control systems in the MPI CyberMotion Simulator	132
	6.4	Generalization of the results	134
	6.5	Recommendations	135

Bibliography   137

Appendices   145

A Procedure for flight test data collection   147

B Measurement setup for data collection   157

C State space identification frequency responses   159

D Experiment briefing   167

E Mathematical proofs for adaptive control design   173

Acknowledgments   175

CHAPTER 1

Introduction

I N recent years, congestion in the transportation system has become
an urgent issue. The increasing road traffic affecting many cities
all over the world has generated several problems in terms of delays,
fuel consumption and pollution, seriously impacting the output of
global economies. Delays due to road congestion have been estimated
to cost approximately €100 billion per year in Europe only [Perfect
et al., 2015b]. These issues have led regulators to consider the implementation of drastic changes in the general public transportation.
A proposed solution has been to reduce congestion by exploiting the
vertical dimension, moving commuting traffic from the ground into
the air. The final result would be the creation af a Personal Aerial
Transportation System (PATS) composed of Personal Aerial Vehicles
(PAVs) that the civil community would use as new means of transport
[Truman and de Graaff, 2007].

This new concept has been investigated in many parts of the world
and is generating radical changes in the transport technology, involving
both fixed- and rotary-wing airframes. The Moller Skycar [Moller,
1998], the NASA Puffin [Moore, 2010] and the Terrafugia TF-X [Wax,
2010], are some examples of small aircrafts that could be used to move
a proportion of road traffic into the air to reduce congestion problems.
Many of these attempts are aircrafts with Vertical Takeoff and Landing
(VTOL) capabilities. This is an important feature as it combines
the benefits of conventional road transportation (e.g. door-to-door,
available to all) and air transportation (e.g. high speed, comparatively
free of congestion) while avoiding the need to build infrastructures
such as airports or roads [Perfect et al., 2014]. Although a considerable
number of prototypes has been proposed for personal transportation
purposes, none has achieved mass production until now. The main
cause lies in the financial capabilities and the piloting skills these
vehicles still require.

To face these issues and to further investigate the possibility of
a transportation system revolution, the European Commission (EC)
funded an out of the box study from 01/2011 to 12/2014, the myCopter
project [Nieuwenhuizen et al., 2011]. During this period, partners from
different European countries identified new concepts for air transport
that could be used to achieve a PAT in the second half of the 21st
century. In particular, it was investigated what features PATS should

present and PAVs should have to be broadly accepted. A key element recognized in the project was the reduction of training costs. Today, the costs associated with learning how to drive a car are approximately of €1.500, while pilots training costs are still of the order of €25.000-40.000 [Bülthoff et al., 2011]. For PAVs to be massively accessible, it would be crucial to have same costs and training duration as those required to obtain a car driving license today [Perfect et al., 2015a]. This could be achieved by equipping PAVs with augmented control systems that would allow the achievement of flying qualities (also called Handling Qualities) considered intuitive and straightforward for novice PAV pilots. In this way, student pilots would be able to develop skills necessary for safe operations in a significantly reduced period of time.

Clearly, autonomous vehicles should be the ultimate goal for PAVs operations. However, the technology required to allow for unmanned aerial systems is not mature enough to be widely deployed. Furthermore, regulations concerning operations of autonomous vehicles need infrastructures that are still not present in non segregated airspace [Alan Simpson and Stoker, 2009].

1.1   Helicopters as Personal Aerial Vehicles

Several studies have been conducted to improve helicopters HQs in the last decades [Tischler et al., 2008]. However, most innovations have been implemented in the military field where progresses concerning helicopter controllability and agility are generally crucial for mission tasks performed by highly trained pilots. The main method used for helicopters HQs assessment has been the ADS-33E-PRF [Baskett, 2000], which is today a standard in the military field.

The ADS-33E-PRF was developed with professional test pilots to assess different helicopter response types while performing specific maneuvers called Mission Task Elements (MTE). Three main response types are considered in the ADS-33E-PRF: the Rate Command (RC), the Attitude Command-Attitude Hold (ACAH) and the Translational Rate Command (TRC) response types. The RC response type is common in many helicopters and is characterized by an angular constant rate in roll, pitch or yaw, due to a step control deflection. Helicopters with a stability control augmentation system (SCAS) have an ACAH response type. In this case, a step control deflection causes angular rates that return to zero after few seconds. Because of the SCAS, the ACAH response type is generally easier to control than the RC response type. Some SCAS can also give a TRC response type, in which control deflections directly command the helicopter linear velocity with respect to the ground. The TRC response is the easiest to control allowing pilots to maintain the hover regime even in degraded visual conditions. In the ADS-33E-PRF three levels are defined to rate the described helicopter responses: Level 1 corresponding to desirable handling characteristics, Level 2 corresponding to acceptable handling characteristics, and Level 3 corresponding to undesirable handling characteristics.

Based on these studies, partners of the myCopter project recognized the importance of defining response types and handling qualities for PAVs pilots. However, the method described in the ADS-33E-PRF was considered inappropriate for PAVs handling qualities assessments, as novice pilots are generally not trained to give specific evaluations on vehicle responses. Therefore, new methods and techniques were implemented to assess candidate vehicle response types and to define HQs for PAVs. In these studies, the helicopter response types described in the ADS-33E-PRF were analyzed to evaluate if they could allow novice pilots to perform flight maneuvers with an acceptable level of precision, in a repeatable manner and in a safe way [Perfect et al., 2015b]. The evaluation was done in simulation by using a PAV model consisting of simplified linear helicopter dynamics and a transfer function representation of the different response types. The response types were manually tuned through experiments performed at University of Liverpool with the use of the HELIFLIGHT-R simulator [Perfect et al., 2013]. Novice pilots were invited to perform piloted closed loop control tasks. These studies showed that the majority of participants could fly TRC response types in hover and low speed flight conditions. The transfer functions parameters of the TRC were adjusted to ensure high performance and low workload for all participants. Therefore, the TRC response type was assessed well suited for PAVs pilots [Perfect et al., 2015a].

The definition of response types and HQs for novice pilots was a crucial achievement of the myCopter project. However, building a new vehicle with such features was not among the project's goals. Therefore, the results obtained at the end of the project left an important unanswered question: is it possible to augment existing vehicles to obtain same performance as those achieved with a PAV model? Among many possible PAV candidates with VTOL capabilities, civil light helicopters represent existing technologies that reflect many properties a PAV should have (e.g. size, weight, number of seats). Furthermore, they have been tested and produced for almost a century. Therefore, considering civil light helicopters as PAVs candidates would represent a valid alternative to the common practice of creating new vehicles with PAVs features.

However, civil helicopters are still today a niche product. Learning how to fly a helicopter is quite challenging and requires time and dedication. Therefore, civil light helicopters are still far from being considered broadly accessible to the general public and many challenges need to be faced to verify whether it is possible to transform them into PAVs. A good knowledge of the state of the art of helicopter system identification and augmentation represents an important starting point to address these challenges. System identification is an engineering tool that provides models well suited for augmented control system applications. Control augmentation is a necessary step to transform civil helicopters dynamics and to achieve required HQs and response types. Identification and augmentation approaches will be considered in the next two sections.

1.2 Helicopter identification

In the past few decades, system identification studies have significantly contributed to the improvement of existing helicopters and the development of new ones. In particular, many dynamic models of large-scale military helicopters have been developed [Baskett, 2000; Fletcher, 1995a; Ham et al., 1995]. In 1995 the AGARD Working Group WG 18 created a large database of flight tests measurements for three different helicopters: the Apache AH-64, the MBB BO-105 and the SA-330 Puma [AGA, 1995]. This database was then used to apply and validate different frequency and time domain system identification methods.

This and similar works on system identification [Hamel and Kaletka, 1997] have allowed a better understanding of helicopter dynamics and the development of new designs for helicopter control systems [Chen and Hindson, 1986; Greiser and Lantzsch, 2013]. Generally, disadvantages of using non-linear full-flight-envelope models are linked to their difficulties at predicting some fundamental dynamics. Conversely, system identification models are built for specific flight conditions and provide better results [Fletcher, 1995b].

Nevertheless, so far the civil light helicopter field has not fully benefited from the advantages system identification methods can offer. The main body of work involves studies conducted in 1993 on a MBB BO-105 helicopter used to investigate higher order models including rotor degrees of freedom [Fu and Kaletka, 1993; Tischler and Cauffman, 1992]. These studies have been continued on the EC-135 from Airbus Helicopters, operated by the German Aerospace Center [Kaletka et al., 2005], and currently focus on developing inverse dynamic models to cancel out inherent helicopter dynamics for simulation purposes [Greiser and von Gruenhagen, 2013]. A recent application of system identification methods on civil light helicopters was the development of a full flight envelope helicopter simulation of the IAI Bell 206, obtained by stitching together dynamic models at different flight conditions [Zivan and Tischler, 2010].

System identification techniques presented in literature have important requirements: the presence of an experienced test pilot able to perform specific maneuvers for identification purposes; a reasonable amount of test flight hours to increase the chance of selecting good trials from the collected data; the availability of software tools developed for time or frequency domain system identification purposes; expert engineers to build reliable models from the data.

These requirements make civil helicopter manufacturers skeptical about including system identification studies into the production cycle. Therefore, the design of light weight helicopters is still done with manual tuning and trial-and-error methods, based on previous experience. An increasing application of system identification techniques would be achieved if the whole process was simplified and reduced and if satisfactory results were obtained without necessarily following all requirements considered in literature.

1.3 Helicopter augmentation

Helicopter dynamics are characterized by complex behaviors that make them difficult to model and control. In the last few decades, a big effort was made to apply control techniques to enhance helicopters stability and controllability and to meet demanding specifications like the HQs requirements defined in the ADS-33E-PRF [Theodore et al., 2014; Tischler et al., 2002]. These studies were motivated by the importance of reducing pilot workload and to improve safety.

Recently, helicopter models obtained through system identification have contributed to the implementation of control augmented systems in simulation. Therefore, costs and safety risks have been reduced by avoiding extensive flight tests for control tuning and validation [Theodore et al., 2014; Walker and Postlethwaite, 1996]. In particular, identified linear models have allowed the implementation of multi inputs-multi outputs robust control techniques, well suited for helicopter dynamics [Postlethwaite et al., 2005; Walker et al., 2000].

However, most of the research on helicopter control has been conducted in the military field, as for the system identification case. Therefore, studies to improve helicopter HQs have been focusing on experienced highly trained pilots. So far, both scientific and civil helicopter communities have not shown any interest in applying control augmented approaches to achieve helicopter dynamics considered intuitive and easy to control for novice pilots.

1.4   Research motivation

The main motivation for this thesis is directly related to the question as to whether civil light helicopters can be considered as possible PAVs candidates. Answering this question presents several challenges.

The first challenge concerns the state of the art of civil helicopters and the lack of research in this field, which make these vehicles far from being associated with PAVs. Although a considerable literature has been devoted to system identification for military purposes, it is not clear if the civil field can afford this kind of research. Some of the reasons why system identification is not considered suitable for civil applications are the need for prior knowledge, experience, expensive instrumentation technologies and multiple flight test hours. Therefore, a further motivation for this thesis would be to investigate whether the civil helicopter community can afford identification studies. One possibility would be to verify if the knowledge and experience gained during the last decades in the military field can be adapted to overcome the civil field limitations.

Another motivation for this thesis is related to the fact that most of the helicopter control literature has been focused on improving handling qualities and response types for professional and highly trained pilots. Understanding whether civil helicopters can be augmented to achieve PAVs handling qualities would mean integrating the main results of the myCopter project with the most recent helicopter control literature in order to steer the research interests towards inexperienced pilots.

1.5   Objectives

Three objectives were formulated in this thesis to investigate whether system identification techniques can be successfully applied in the civil helicopter field and to prove whether civil helicopters dynamics can be augmented to achieve PAVs response types and HQs.

Thesis objectives

1. In the first objective, a Robinson R44 Raven II helicopter is identified. The aim is to investigate whether the current helicopter identification literature allows for reliable models to be implemented despite the civil field limitations. The hover condition is taken into account as it is commonly considered one of the most difficult to perform for novice pilots. Finally, the MPI Cyber-Motion Simulator [Nieuwenhuizen and Bülthoff, 2013] is used to validate the reliability of the identified model by performing specific control tasks with an experienced helicopter pilot in a simulated virtual environment.

2. In the second objective, robust control augmentation techniques (H∞ and μ-synthesis) are applied to the helicopter model identified in the first objective. In this way, it is possible to investigate whether civil helicopters dynamics and HQs can be modified to achieve those defined for PAVs.

3. In the third objective, the implemented augmented systems are evaluated by performing an experiment in the MPI CyberMotion Simulator. In this experiment, subjects without any prior flight experience are invited to perform piloted closed loop control tasks with the augmented systems, with the identified helicopter model and with the PAV reference model. Then, comparisons are made in terms of subjective and objective performance and workload. The aim of the experiment is to provide insights into the augmented systems features and to verify whether civil helicopters can be augmented to achieve performance and workload levels defined for PAVs.

1.6 Approach and outline of the thesis

To achieve the defined objectives, an approach is proposed and presented in five main chapters. Chapter 2 tackles the first objective of the thesis, i.e. the identification of a civil light helicopter. A frequency based system identification method is implemented to identify a linearized model of a Robinson R44 Raven II in hover. First, a lean and practical procedure is presented that provides a reliable collection of control inputs and helicopter responses for system identification purposes. Then, the adopted system identification method is introduced. This method is based on the theory presented in literature by Tischler [Tischler and Remple, 2012]. The first step is the implementation of a non-parametric model followed by a transfer function model. These two models are useful to assess which dynamic characteristics are captured in the collected data and are to be included in the final state-space model. Then, a state-space fully-coupled model is identified and validated. The validation is done in the time domain by using different input signals than those considered during the identification to evaluate the model's predictive capabilities. Finally, the identified model is assessed by a helicopter pilot while performing hover and low speed control tasks in the MPI CyberMotion Simulator. Both time domain validation and helicopter pilot evaluation provide important insight into the reliability of the identified model.

Chapter 3 presents the second objective of the thesis, i.e. the implementation of robust control techniques applied to the identified helicopter model of Chapter 2. The H∞ control method is chosen because of its capability to deal with Multi-Input Multi-Output (MIMO) linear systems subject to uncertainties and external disturbances [Walker et al., 2000]. Furthermore, its frequency based approach allows for PAVs HQs specifications to be easily included. A μ-synthesis robust control technique is also implemented and compared with the H∞ approach. The μ-synthesis includes model uncertainties into the optimization problem to achieve robustness, a feature not guaranteed by the H∞ control design [Zhou and Doyle, 1999]. Moreover, it is a frequency based method as well as the H∞ one, thus presents all the advantages described before. A disadvantage is that resulting controllers have typically high orders. In this chapter H∞ and μ-synthesis robust control techniques are compared against stability and performance requirements. The robustness achieved with the two control techniques is tested against parametric uncertainties and external disturbances. The external disturbances are chosen as real atmospheric turbulences that might be experienced in hover and low speed flight regimes. The parametric uncertainties are associated with the identified parameters of the helicopter model. A problem shown in this chapter is that both implemented control techniques cannot ensure required performance against model parametric uncertainties. This problem is faced in Chapter 5.

Chapter 4 presents the final objective of the thesis, i.e. comparing the implemented control augmented systems of Chapter 3 with respect to the original helicopter model and evaluating possible discrepancies with respect to each other and with respect to the PAV reference model. To do this, an experiment is conducted in the MPI CyberMotion Simulator. In this experiment, inexperienced pilots are invited to perform piloted closed-loop control tasks. Two maneuvers or Mission Task Elements (MTE) are considered as defined in the ADS-33E-PRF: the Hover MTE and the Lateral Reposition MTE [Baskett, 2000]. Each participant is asked to perform both MTEs with one of the four dynamics: identified helicopter model, H∞ augmented system, μ-synthesis augmented system and myCopter PAV model. The four dynamics are compared in terms of performance and workload in both MTEs. Insight into the main features of the augmented system is gained from the experimental results. A discussion is provided at the end of the chapter that draws some conclusions on the possibility of using civil light helicopters as PAVs candidates.

Chapter 5 proposes a solution to reduce the detrimental effects of model parametric uncertainties on the performance of the augmented systems presented in Chapter 3. An $\mathcal{L}_1$-based adaptive control is considered for this purpose. To verify the effectiveness of this approach, the adaptation is applied to a simple PID-based control. First, the PID-based controller is implemented to augment the nominal helicopter model without uncertainties. Then, an $\mathcal{L}_1$ adaptive controller is designed to restore the nominal responses of the augmented helicopter when parametric uncertainties are included. Finally, a Monte-Carlo study is performed to validate the proposed control architecture against different realizations of the uncertain parameters. Conclusions are given at the end of the chapter about the efficacy of the implemented adaptive control method at reducing the detrimental effect of model parametric uncertainties.

Chapter 6 highlights final conclusions, contributions and findings of each chapter. Furthermore, recommendations about possible future works are given.

1.7 Thesis scope

The work presented in this thesis is based on specific assumptions that bind the validity of the results. An assumption of this thesis is that hover and low speed flight conditions are the most difficult to perform for inexperienced pilots. This is due to the fact that helicopters in these conditions are unstable, non-minimum phase and highly coupled. This assumption leads to the conclusion that answering to the main question as to whether non expert pilots can stabilize and fly an augmented helicopter in hover and low speed flight regime can provide insight into other flight regimes, e.g. high speed maneuvers.

Another assumption considered here is that robust control techniques are particularly well suited for achieving the goal proposed in this thesis. This is supported by the fact that these control methods can easily deal with MIMO linear systems subject to uncertainties and external disturbances or sensor noise. Furthermore, they are frequency based, which allows for a direct integration of the HQs specifications considered in this thesis. However, comparisons with other control techniques (e.g. PID, inverse control dynamics) are not considered here since beyond the scope of the thesis. Further studies could be considered to investigate whether other control techniques can achieve similar results.

Finally, the implemented augmented systems are tested and compared in the MPI CyberMotion simulator by considering a virtual environment in calm air and good external visual conditions. This choice is based on the assumption that a clear assessment of the considered augmented models can be obtained if external factors that could influence this evaluation are not included. However, disturbances, noise or degraded visual conditions are all part of a realistic environment. Although an investigation of these aspects is beyond the thesis objectives, additional experiments might be considered to analyze how these factors can influence inexperienced pilots workload and performance.

1.8 Publications

Most chapters of this thesis have been published or submitted as papers. The notation and the style have been adapted were possible to be consistent throughout the thesis. An overview of publications used in this thesis is given below:

- Chapter 2 is based on the following papers:

  Geluardi, S., Nieuwenhuizen, F. M., Pollini, L. and Bülthoff, H. H., "Data collection for developing a dynamic model of a light helicopter", *Proceedings of the 39th European Rotorcraft Forum, Moscow*, Sep. 2013.

  Geluardi, S., Nieuwenhuizen, F. M., Pollini, L. and Bülthoff, H. H., "Frequency Domain System Identification of a Light Helicopter in Hover", *Proceedings of the American Helicopter Society 70th Annual Forum, Montreal, Canada*, May 2014.

  Geluardi, S., Nieuwenhuizen, F. M., Pollini, L. and Bülthoff, H. H., "State Space System Identification of a Civil Light Helicopter in Hover", *Journal of the American Helicopter Society*, in preparation.

- Chapter 3 is based on the following paper:

  Geluardi, S., Nieuwenhuizen, F. M., Pollini, L. and Bülthoff, H. H., "Augmented Systems for a Personal Aerial Vehicle Using a Civil Light Helicopter Model", *Proceedings of the American Helicopter Society 71st Annual Forum, Virginia Beach, USA*, May 2015.

- Chapter 4 is based on the following paper:

Geluardi, S., Venrooij, J., Olivari, M., Pollini, L. and Bülthoff, H. H., "Validation of Augmented Civil Helicopters in the MPI CyberMotion Simulator", *Journal of Guidance, Control, and Dynamics*, in preparation.

- Chapter 5 is based on the following paper:

  Picardi, G., Geluardi, S., Olivari, M., Pollini, L., Innocenti, M. and Bülthoff, H. H., "$\mathcal{L}_1$-based Model Following Control of an Identified Helicopter Model in Hover", *Proceedings of the American Helicopter Society 72th Annual Forum, Palm Beach, Florida*, May, 2016.

The following papers have also been produced during the course of the thesis work, but are not included in this thesis:

- Gerboni, C. A., Geluardi, S., Olivari, M., Nieuwenhuizen, F. M., Bülthoff, H. H. and Pollini, L., "Development of a 6 dof nonlinear helicopter model for the MPI Cybermotion Simulator", *Proceedings of the 40th European Rotorcraft Forum, Southampton*, Sep. 2014.

- D'Intino, G., Olivari, M., Geluardi, S., Venrooij, J., Innocenti, M., Bülthoff, H. H. and Pollini, L., "Evaluation of Haptic Support System for Training Purposes in a Tracking Task", *Proceedings of the SMC 2016 IEEE International Conference on Systems, Man, and Cybernetics*, submitted.

- Olivari, M., Geluardi, S., Venrooij, J., Nieuwenhuizen, F. M., Pollini, L., and Bülthoff, H. H. "Identifying Time-Varying Neuromuscular Response: Regularized Recursive Least-Squares Algorithms", *IEEE Transactions on Cybernetics*, in preparation.

CHAPTER 2

Identification of a civil light helicopter in hover

This chapter presents the identification of a Multi-Input Multi-Output fully coupled model of a civil light helicopter in hover. A frequency domain identification method is implemented for the identification. It is discussed that the chosen frequency range of excitation captures some important rotor dynamic modes. Therefore, studies that require coupled rotor/body models are possible. A parametric transfer function identification is first implemented to provide useful information for selecting an appropriate state-space model formulation. A state-space model is then implemented. The model is validated in the time domain with different input signals than those used during the identification. Good predictive capabilities are achieved and high frequency rotor-body dynamics are captured. Therefore, the model is well suited for handling qualities studies and control system designs. A further validation of the model is made by performing specific maneuvers with a helicopter pilot in the CyberMotion Simulator.

The contents of this chapter are based on:

Paper title	Data Collection for Developing a Dynamic Model of a Light Helicopter
Authors	S. Geluardi, F. M. Nieuwenhuizen, L. Pollini and H. H. Bülthoff
Published in	Proceedings of the 39th European Rotorcraft Forum, Moscow, Russia, 2013

- and -

Paper title	Frequency Domain System Identification of a Light Helicopter in Hover
Authors	S. Geluardi, F. M. Nieuwenhuizen, L. Pollini and H. H. Bülthoff
Published in	Proceedings of the American Helicopter Society 70th Annual Forum, Montreal, Canada, 2014

- and -

Paper title	State Space System Identification of a Civil Light Helicopter in Hover
Authors	S. Geluardi, F. M. Nieuwenhuizen, L. Pollini and H. H. Bülthoff
In preparation	Journal of the American Helicopter Society

IN this thesis, the implementation of a system identification model
was preferred to the creation of a non-linear full-flight-envelope
model. This choice relies on previous studies, which demonstrated
the deficiencies of complex non-linear model in predicting some fundamental dynamics [Fletcher, 1995a]. Indeed, dynamics like primary
roll, vertical response or pitch/roll cross-coupling may not be correctly captured when the model is implemented to be valid over the
full-flight-envelope. Therefore, models identified for specific flight
conditions are generally more reliable.

System identification consists of a sequence of specific steps that
allow for the implementation of a model of a physical system from
measured test data. Nowadays, it is an established routine procedure
in the fixed wing aircraft field as it is used to obtain 3 or 6 Degrees of
Freedom (DoF) linearized rigid body models [Klein and Morelli, 2006].
In the last decades, a big effort has been made to apply identification
methods in the rotorcraft field as well [Hamel and Kaletka, 1997]. In
particular, the AGARD Working Group WG 18 on 'Rotorcraft System
Identification' aimed to investigate how identification theories can be
applied to rotorcraft systems. The result was the creation of a large
database of flight tests measurements for three different helicopters:
the Apache AH-64, the MBB BO-105 and the SA-330 Puma [AGA,
1995]. This database was then used to apply and validate different
frequency and time domain system identification methods. This and
similar works on system identification [Hamel and Kaletka, 1997] have
opened a new branch of research focused on better understanding
helicopter dynamics and developing new helicopter control systems
[Chen and Hindson, 1986; Greiser and Lantzsch, 2013]. Nevertheless,
the knowledge and experience gained from this kind of research has
mainly been taken into consideration in the military field [Ivler and
Tischler, 2013; Jategaonkar et al., 2004; Tischler and Remple, 2006].
As a result of this, many studies have been conducted in the past years
on large-scale military helicopters [Baskett, 2000; Fletcher, 1995a;
Ham et al., 1995].

Today, system identification is sufficiently mature to allow applications on novel types of vehicles, e.g. Unmanned Aerial Vehicles (UAVs)
[Dorobantu et al., 2013]. The reason is linked to the short design

cycles and the tight budgets involved in this fields that do not always allow accurate and reliable analytical models to be implemented [Theodore et al., 2004].

However, the civil light helicopter field has not yet fully benefited from the advantages system identification methods can offer. One reason is the incapability of affording expensive instrumentation technologies that are commonly used in the military field [Dorobantu et al., 2013]. Another factor linked to costs is the unavailability of multiple hours of flight test, generally needed to increase the probability of collecting reliable measurements. Therefore, so far civil helicopter producers have not shown any interest in system identification studies. As a result, design and development of light weight helicopters are still made with manual tuning and trial-and-error methods, based on previous experience.

This explains why a few system identification studies have been performed on civil helicopters until now. The main body of work involves the research conducted on a MBB BO-105 helicopter to investigate higher order models that can include rotor degrees of freedom [Fu and Kaletka, 1993; Tischler and Cauffman, 1992]. These studies have been continued on the EC-135 from Airbus Helicopters, operated by the German Aerospace Center [Kaletka et al., 2005], and currently focus on developing inverse dynamic models to cancel out inherent helicopter dynamics for simulation purposes [Greiser and von Gruenhagen, 2013]. A recent application of system identification methods on civil light helicopters has been the development of a full flight envelope helicopter simulation of the IAI Bell 206, obtained by stitching together dynamic models at different flight conditions [Zivan and Tischler, 2010].

The use of system identification would increase in the civil field if the whole process was simplified and reduced and if satisfactory results were obtained without following all requirements considered in literature. This chapter focuses on testing whether system identification methods are mature enough to be easily implemented in the civil field. In particular, a simplified and lean identification procedure will be proposed to overcome some of the limitations the civil helicopter field presents. The system identification described in this chapter represents the first step towards the thesis goal of augmenting a civil light helicopter model to achieve dynamical characteristics and handling qualities of a PAV.

The goal of this chapter is the identification of a Robinson R-44 Raven II helicopter (Figure 2.1) in hover condition. This helicopter is light weight and possesses a single engine, a semi-rigid two-bladed main rotor and a two-bladed tail rotor [Robinson, 1992]. The Robinson R-44 was selected in this thesis as it presents comparable features as those of a PAV (e.g. size, weight, number of seats) and is commonly employed in the civil field for non expert pilots training. Therefore, it was considered well suited for the purpose of the thesis. In this thesis, the system identification of the model is developed for the hover condition as the high task bandwidth of this maneuver makes it one of the most complex to perform for pilots with limited flight experience.

Figure 2.1: Robinson R44 helicopter

An important aspect analyzed in this chapter is the choice of the model dynamic complexity. A 6 DOF model is generally adequate for handling qualities evaluations. However, higher order model structures are necessary for flight control system design or for model validation studies [Hamel and Kaletka, 1997]. Many works demonstrated that high bandwidth control systems for helicopters need to include rotor degrees of freedom. In [Chen and Hindson, 1986] a variable-stability CH47 helicopter was used to demonstrate how rotor dynamics and lags in the control system can influence the feedback gain limits. Later, Tischler investigated high order mathematical helicopter models and proved that for a hingeless single-rotor helicopter the coupled body/rotor-flapping mode can limit the gain on attitude control feedback, while the lead-lag mode can limit the gain on attitude-rate control feedback [Tischler, 1991]. Recent studies by DLR in Germany have confirmed the importance of suppressing the air resonance due to the regressive lead-lag mode [Greiser and Lantzsch, 2013].

These studies suggest that considering rotor's DOF and in particular lead-lag modes can be necessary to implement reliable augmented control systems and to analyze their differences. Since this represents the second objective of the thesis, an analysis on model complexity will be considered in this chapter to assess whether the collected data are able to capture rotor/body coupling dynamics.

The chapter is structured as follows. First, a non-parametric system identification is considered based on the collected data. Then, a parametric transfer function identification is implemented to provide useful information for determining an appropriate state-space model formulation. Finally, a parametric state-space model is identified. The model is first validated in the time domain to verify its predictive capabilities. Then, a final assessment is made with a helicopter pilot through piloted closed-loop control tasks performed in the MPI CyberMotion Simulator.

2.1   Collection of flight test data

A crucial step of the system identification process is the data collection. Having reliable data is necessary to produce a final model close to the real physical system. Identifying system dynamic characteristics of interest (i.e. the modes of the system) is not possible if the collected data do not contain this information [Tischler and Remple, 2012].

Three main steps need to be considered to ensure the collection phase provides data sufficiently reliable for system identification [AGA, 1995; Williams et al., 1995]. The first step involves the implementation of a measurement setup, which includes the choice of all sensors necessary for measuring pilot inputs and helicopter response.

The second step concerns the flight maneuvers choice. Piloted frequency sweeps are usually selected to allow for frequency domain identification methods to be implemented. Doublets maneuvers are generally chosen to validate the identified model in the time domain, using different inputs than those used during the identification.

The third step consists of performing flight tests for collecting data. In this phase, pilot inputs and helicopter responses are measured, while a helicopter pilot performs the selected maneuvers per each control axis.

These three steps will be now considered in details.

2.2   Development of the measurement setup

This section focuses on the development of the measurement setup for collecting inputs and outputs of the helicopter model to identify. First, the selected signals are described. Then, the instrumentation is presented and particular attention is devoted to the validation of the proposed setup for the collection of the model inputs.

2.2.1   Required signals

Selecting the signals to be measured is directly related to the applications of the model to identify. In this thesis an important application is the design of control augmentation techniques, which usually requires rotor/body coupling dynamics to be captured. To allow this, the following model outputs were considered: linear accelerations, angular rates, linear velocities and attitude. The signals selected as model inputs were measured as pilot controls deflections.

2.2.2   Instrumentations for the output vehicle signals

The instrumentation for collecting the helicopter output signals was composed of an Inertial Measurement Unit (IMU) and two Global Positioning System (GPS) antennas. The choice of using two GPSs was made to reduce ionospheric errors by modeling and combining satellite observations on two different frequencies.

The IMU comprised Fiber Optic Gyros (FOG) and Micro Electrical Mechanical System (MEMS) accelerometers. The accuracy of the two GPS antennas coupled with the stability of the IMU measurements provided a stable and continuous 3D navigation solution, even through periods when satellite signals were not available. To enhance this function, the position of the GPS antennas with respect to the IMU was accurately measured. During the data collection the two GPS antennas were installed on the left skid of the helicopter while the IMU was located on the helicopter ground behind the front left seat (Figure 2.2). The longitudinal and lateral location of the CoG was determined by measuring weight and position of instrumentation and people inside the helicopter during the flight tests. The longitudinal and lateral relative position of the IMU with respect to the CoG was computed to obtain physically coherent vehicle data. The vertical CoG location was not directly measured and, therefore, left as unknown parameter to identify. The signals obtained from GPS and IMU were: inertial helicopter position (x, y, z), attitude (ϕ, θ, ψ), angular rates (p, q, r) and linear accelerations (a_x, a_y, a_z). The selected sample rate was 100 Hz. The schematic representation of the setup is given in Figure 2.3. Details of the sensors used in the measurement setup are provided in Appendix B.

2.2.3   Instrumentation for piloted control inputs

Measurements of the control displacements should be performed without affecting the pilot. For this reason, optical sensors were employed

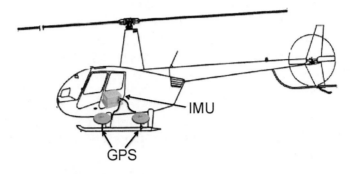

Figure 2.2: IMU and GPS antennas positions

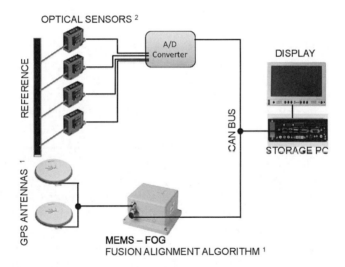

Figure 2.3: Schematic overview of the measurement setup used for the
flight tests.

[1]http://www.novatel.com/products/span-gnss-inertial-systems/span-combined-systems/span-cpt

[2]http://www.wenglor.com

capable of measuring distances without mechanical contacts. Four optical sensors were placed on the pilot controls to measure the control inputs: longitudinal cyclic stick deflection (δ_{long}), lateral cyclic stick deflection (δ_{lat}), collective lever deflection (δ_{col}) and pedals deflection (δ_{ped}). Furthermore, all sensors were attached to the controls on the left side of the helicopter, having the pilot sitting on the seat on the right side.

Optical sensors can measure a linear distance from a specific reference object. In the considered setup, the sensors were rigidly attached to the controls, while flat surface references were located at specific distances. In this way, a continuous measurement was obtained of the distance between the point on the pilot control, in which the sensor was attached, and the surface used as reference. However, helicopter control deflections are better expressed as angles, due to the type of motion allowed. Therefore, a mapping was defined between the measured linear distances and the relative angular deflections. To ensure a correct mapping, different scenarios of the measurement setup location were analyzed in simulation.

A possible scenario is presented in Figure 2.4. Here, the optical sensor is attached to the cyclic stick. Three different stick positions are considered: the centered and the two extreme positions (angular displacement α). The variables shown in the figure are: the distance of the sensor with respect to the cyclic hinge (l), the distance of the reference plate with respect to the cyclic hinge (d), and the height of the reference plate (h). By changing any of these variables a different relationship is obtained between the measured distance (x) and the angular displacement (α). As shown in Figure 2.5, the slope of the plate (ϕ) plays an important role in determining an univocal mapping between the measured distance x and the angular displacement α.

In particular it can be noticed that ambiguous results might be obtained for specific configurations. In this case, multiple angular displacements (α_1, α_2) are associated to the same measured distance (x). With this kind of analysis, a correct configuration was found free from ambiguous relationships between measured distances and angular displacements. Although this was verified in simulation, a practical application resulted problematic due to the difficulty of precisely measuring all variables shown in Figure 2.5. Therefore, an empirical

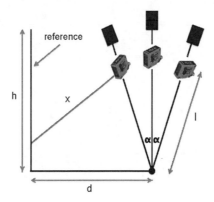

Figure 2.4: Schematic representation of the relation between the linear
distance measurement(x) and the angular displacement (α)

method was implemented to validate the selected configuration. The
empirical method was applied to each control axis and is described in
the following section.

2.2.4 Validation of the measurement setup for piloted control inputs

The actual relationship between the measured distance x and the
angular displacement of the control stick α, was found generating a
look-up table. For each control axis, distances x and angles α were
measured in several control configurations. Then, the measurements
were interpolated to find a final relationship (Figure 2.6). As can be
seen for the longitudinal axis of the cyclic stick, a univocal mapping
is obtained between the measured distance (x) and the measured
angle (α). This validates the correctness of the configuration selected
in simulation by applying the analytical method described in the
previous section. The same result was obtained for all control axes.

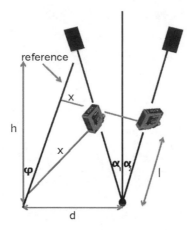

Figure 2.5: Bad configuration

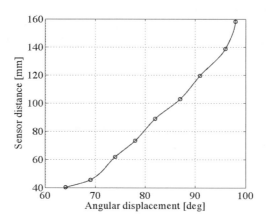

Figure 2.6: Relationship between the cyclic longitudinal angular displacement (α) and the linear distance (x)

2.3 Choice of flight maneuvers

Two kind of maneuvers were selected to be performed during the test
flights: piloted frequency sweeps and doublets. Piloted frequency
sweeps are suited for frequency domain identification method applications. Doublets are generally used to validate the reliability of the
identified model, when other kinds of maneuvers (e.g. Frequency
sweeps) are used during the identification process. Furthermore, due
to their simplicity, doublets are particularly suited to train pilots at
the beginning of the flight test. Moreover, their simple form allows the
identification of main aircraft dynamic characteristics and the analysis
of data consistency. An important feature of these maneuvers is that
they are symmetrical. This allows the restriction of the helicopter
dynamics in the range of transients where the model is expected to
be valid [Tischler and Remple, 2012].

The frequency sweeps were designed with a range of excitation
between 0.05 and 2.5 Hz ($\approx$ 0.3 - 16 rad/s), to allow for rotor-body
coupling modes to be captured.

Both frequency sweeps and doublets where designed to generate a
change in vehicle attitude between ± 10 and ± 20 degrees and a change
in velocity of about ± 5 m/s. Generally, maneuvers with a wider
displacements lead to big drifts from the trim condition and should be
avoided. On the other hand, smaller control amplitudes could cause
a too low signal to noise ratio in the measurements [Williams et al.,
1995].

Because of these strict guidelines, a preliminary pilot training was
considered necessary.

2.3.1 Pilot training phase

The flight condition of interest in this thesis is hover. Many helicopters
show strong coupled degrees of freedom and are highly unstable under
this condition. For these reasons a preliminary training phase was
considered necessary, on earth and in flight, to make the pilot capable
of performing the chosen maneuvers safely. This, while ensuring
measurements sufficiently reliable for the identification process.

The following training phase was considered. First, a verbal description of the specific maneuvers was given to the pilot. Then, the pilot practiced at performing the required maneuvers on the ground till correct magnitudes and input timings were achieved. During the ground training the pilot was provided with a monitor to watch in real-time the maneuvers he was performing. Finally, a few more trials were performed in flight before the actual flight test. The training phase was particularly important as the pilot had never performed this kind of maneuvers before.

2.4   Flight tests

The data for system identification were collected during two flight tests, each with a duration of one hour. Several piloted frequency sweeps and doubles were recorded for each control axis at an altitude of 10 meters, hence in ground effect. The weather conditions were good with a medium temperature of 25 degrees Celsius and wind velocities less than 5 knots. The measurement setup was placed inside the helicopter so as to allow the presence of a flight test engineer on board, responsible for calling the maneuvers and monitoring the instrumentation.

In Figure 2.7 the recordings of two concatenated frequency sweeps are shown for the longitudinal axis. Each frequency sweep has a duration of about 100 seconds. The sweep maneuvers start in hover with a few seconds of trim and end with the same initial trim condition. The longest period of the sweeps is about 20 s which corresponds to a frequency of about 0.05 Hz. Then the pilot slowly increases the frequency of the sweep till a period of 0.4 s is reached (≈ 2.5 Hz). Figure 2.7 shows also the primary helicopter responses to the longitudinal cyclic stick deflection. The control input is given in degrees, having mapped the measured data into linear angular displacements. The primary vehicle responses due to a longitudinal cyclic input are the linear velocity (u), the pitch rate (q) and the pitch angle (θ). It can be noticed that the pitch rate remains within ± 25 deg/sec. This size of excitation ensures the identification of models that are accurate for maneuvers with large excursions. Furthermore, it is important

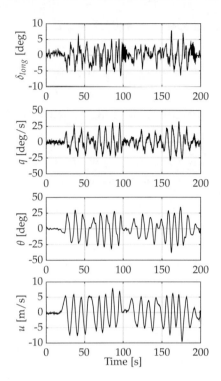

Figure 2.7: A frequency sweep in the longitudinal axis in hover. $\delta_{lon} =$ cyclic longitudinal deflection, $u =$ longitudinal translational velocity, $\theta =$ pitch angle, $q =$ pitch rate.

to notice that the high frequency content of the pitch rate (q) is not present in the longitudinal translational velocity (u) and the pitch angle (θ), where only low frequencies are involved. High frequency content is important for analyses of model complexity that involve rotor/body couplings. Therefore, the pitch rate response q/δ_{lon} will be considered in detail in the next identification phases presented in this chapter.

The concatenation of two or more frequency sweeps as seen in Figure 2.7 allows for rich spectral content to be obtained over the frequency range of interest. Therefore, the same procedure was applied to each control axis. The data collected during the flight tests validated the measurement setup and the adopted method for data collection.

The pilot was able to correctly perform the selected maneuvers without being influenced by the sensors attached to the controls.

2.5 Frequency domain identification method

The frequency domain system identification method developed by Tischler [Tischler and Remple, 2012] was implemented in this thesis using the numerical computation environment MATLAB®. This method was selected because of the several advantages it presents for identifying helicopter models. Non-parametric as well as parametric identification models can be implemented with the possibility to eliminate bias effects and noise in response measurements. Furthermore, tools like coherence functions, composite windowing and frequency response conditioning can be used to increase the reliability of identified models. The possibility of precisely determine time delays is also allowed.

As previously discussed, the applications considered in this thesis (flight control system design and handling qualities evaluation) require higher order model structures that include rotors degrees of freedom [Ivler and Tischler, 2013; Tischler, 1991]. Therefore, a rotor/body coupling high order model was considered. In the next sections a brief overview will be given of the theory behind the frequency domain identification method adopted in this thesis. Then, results will be presented, obtained by applying the identification method on the collected data.

2.5.1 Non-parametric identification

The first step of the system identification method proposed in [Tischler and Remple, 2012] consists of computing the entire set of non-parametric single-input single-output frequency responses. As shown in Figure 2.7, two or three time histories are selected for each input-output pair and concatenated such that a richer spectral content is obtained within the frequency range of interest. The input-output cross-spectrum $\hat{G}_{xy}(f)$ and the input autospectrum $\hat{G}_{xx}(f)$ estimates

can then be used to compute the entire set of Single-Input Single-
Output (SISO) frequency response estimates:

$$\hat{H}(f) = \frac{\hat{G}_{xy}(f)}{\hat{G}_{xx}(f)}$$

2.1

An important tool in this method is the coherence function [Bendat
and Piersol, 2010]:

$$\hat{\gamma}_{xy}^2(f) = \frac{|\hat{G}_{xy}(f)|^2}{|\hat{G}_{xx}(f)||\hat{G}_{yy}(f)|}$$

2.2

Coherence function values vary between 0 and 1. A coherence function
decrease can be associated with noise, nonlinearities, lack of input
excitation or lack of rotorcraft response. In general, coherence values
of 0.6 and above are satisfactory for system identification purposes
[Sahai et al., 1999].

Another aspect to take into account is that estimated frequency
responses obtained from helicopter flight data are generally influenced
by partially correlated inputs and input-output couplings. These can
be removed by conditioning the frequency responses and the related
coherence functions. To do so, the conditioned Multi-Input Single-
Output (MISO) frequency responses [Bendat and Piersol, 2010] are
computed for each output as:

$$\hat{\mathbf{H}}(f) = \hat{\mathbf{G}}_{xx}^{-1}(f)\hat{\mathbf{G}}_{xy}(f)$$

2.3

where $\hat{\mathbf{G}}_{xy}$ is the vector of SISO cross-spectra between each input and
a specific output, and $\hat{\mathbf{G}}_{xx}$ is the matrix containing the inputs auto-
and cross-spectra. The Multi-Input Multi-Output (MIMO) frequency
response matrix can be obtained by first computing equation 2.3 for
each output and then selecting the frequency responses associated
to the primary inputs. This selection is generally done to ensure
high coherences. The partial coherence function associated with each
conditioned response is defined as:

$$\hat{\gamma}_{x_i y \cdot (n_c-1)!}^2(f) = \frac{|\hat{G}_{x_i y \cdot (n_c-1)!}(f)|^2}{|\hat{G}_{x_i x_i \cdot (n_c-1)!}(f)||\hat{G}_{yy \cdot (n_c-1)!}(f)|}$$

2.4

where n_c is the number of piloted control inputs and $\hat{G}_{x_i y \cdot (n_c - 1)!}$ indicates the cross-spectrum estimate between the control input x_i and the output y, conditioned by the other $n_c - 1$ control inputs. Coherently, $\hat{G}_{x_i x_i \cdot (n_c - 1)!}$ and $\hat{G}_{yy \cdot (n_c - 1)!}$ respectively represent the input and output auto-spectrum estimates conditioned by the other control inputs.

The computation of the MIMO frequency response matrix and of the partial coherence functions represents the last step of the non-parametric model identification.

2.5.2   Transfer function and state-space modeling

The non-parametric model described in the last section represents a starting point for identifying two possible kinds of parametric models: the transfer function and the state-space models. Implementing a transfer function model provides useful insight into the order of the system, the level of input-output couplings and fundamental dynamic characteristics of the identified system. Furthermore, initial values of the state-space model parameters can be determined.

Transfer functions models are estimated by fitting the individual conditioned frequency responses computed during the non-parametric identification. The fitting is obtained by minimizing the following cost function [Tischler and Remple, 2012]:

$$J_{SISO} = \frac{20}{n_\omega} \sum_{\omega_1}^{\omega_{n_\omega}} W_\gamma \left[W_g \left(|\hat{T}_c| - |T| \right)^2 + W_p \left(\angle \hat{T}_c - \angle T \right)^2 \right] \quad \boxed{2.5}$$

where $\hat{T}_c$ is the conditioned frequency response estimate obtained from the collected data and T is the selected transfer function model to identify. The frequency points are generally spaced uniformly over a log-frequency scale and the number of frequency points n_ω is usually fixed to 20 [Tischler and Remple, 2012]. In this way the gain in Eq. 2.5 reduces to 1. Finally, the frequency range of interest is selected between ω_1 and ω_{n_ω}.

In this cost function, the response magnitudes and phases are considered separately and scaled with W_g and W_p, respectively [Tischler and Remple, 2012]. The choice of W_g and W_p depends on the units of measure used for magnitude and phase. The weighting function W_γ depends on the value of the coherence function at each frequency point.

Identified transfer function models with associated cost function values $J \leq 100$ have good levels of accuracy and can be used in many key applications, such as handling qualities analysis and control system design [Ham et al., 1995].

State-space models, on the other hand, are better suited for applications that require coupled MIMO behavior (e.g. model validation in simulation and high frequency control system design) as they allow for a simultaneous fitting of the different SISO responses while ensuring coupling constrains. A state-space model is obtained by truncating the Taylor series of the general helicopter dynamic equations at the first order. Furthermore, rotor dynamic equations can be added to account for high frequency modal effects. The result is a hybrid formulation of the linearized model in which stability and control derivatives are the unknown that need to be estimated [Tischler and Cauffman, 1992].

In the state-space model identification, the minimization is performed by summing the n_{tf} input-output cost functions as defined for the SISO case, see Eq. 2.5:

$$J_{MIMO} = \sum_{l=1}^{n_{tf}} J_{SISO,l} \qquad \boxed{2.6}$$

Generally, each cost function minimization is considered adequate when $J_{SISO,l} \leq 200$. The minimization of the state-space model is considered successful when $J_{MIMO}/n_{tf} \leq 100$. The predictive capability of the estimated state-space model is usually assessed with a time domain cross-validation by considering flight test data independent from those used during the identification phase [Tischler and Remple, 2012].

The following section presents a brief description of the data collected during the flight tests. Then, results will be shown obtained by applying the identification theory described above.

2.6   Non-parametric identification results

The concatenated signals obtained from the collected data were used
to create an entire set of nonparametric single-input single-output
frequency responses computed with Eq. 2.3. An example is presented
in Figure 2.8a. Here, the bode plot is shown of the pitch rate response
due to the longitudinal stick deflection, conditioned by the other inputs
$(q/\delta_{lon} \cdot \delta_{lat}, \delta_{col}, \delta_{ped})$. The figure shows a comparison between three

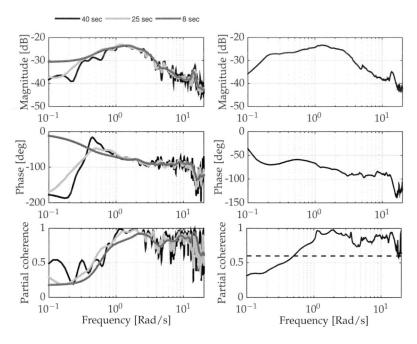

(a) Different windows length   **(b)** Windowing method applied

Figure 2.8: Pitch axis frequency response q/δ_{lon} composite windowing
method. The dashed line represents a coherence value of 0.6. Values of
coherence above this value are considered good.

responses obtained with different window lengths and a 50 percent
window overlap. Using overlapping windows reduces the random error
effects in the spectral estimates [Tischler and Remple, 2012]. It can

be noticed that the 40 seconds window reduces these effects at low frequencies as here the coherence reaches high values. However, in the mid- and high-frequency range the response starts to oscillate. In the mid-frequency range the 25 seconds window gives the best result but the response again has high oscillations for higher frequencies. The 8 seconds window gives poor results at low frequencies (low partial coherence function value) but filters the random error effects at mid- and high- frequencies. Similar results were found in the other input-output pairs. Generally, selecting a window able to provide optimal results over the entire frequency range of interest is very difficult [Tischler and Remple, 2012]. The composite windowing technique described in [Tischler and Remple, 2012] solves this issue by merging the responses obtained with different windows into one response estimate with high accuracy over the entire frequency range of interest. Therefore, the composite windowing technique was applied [Tischler and Cauffman, 1992] to guarantee low-frequency accuracy while reducing the random error effect at high frequencies. Five different window lengths were combined. A window of 40 seconds and one of 8 seconds were selected as the largest and the smallest one, respectively. Three more windows were evenly distributed between these two. The resulting composite responses presented good level of coherence over the entire frequency range of interest (0.3-16 rad/sec) as shown in Figure 2.8b for the pitch-rate response q/δ_{lon}. Here the partial coherence function presents a value above the guideline limit of 0.6 (dashed line in Figure 2.8b) over the entire frequency range of interest.

The non-parametric identification phase showed important dynamic characteristics captured in the collected data. The results of the composite windowing showed some important dynamic characteristics captured in the collected data. For the pitch-rate response q/δ_{long} in Figure 2.8b the peak at around 1.5 rad/sec can be associated to the short period, while the resonance at 14 rad/sec is due to the lightly damped regressive lead-lag mode.

Another interesting response, shown in Figure 2.9, is the vertical acceleration due to the collective deflection, conditioned by the other inputs $a_z/\delta_{col} \cdot \delta_{lat}, \delta_{lon}, \delta_{ped}$. Here, the composite windowing method has already been applied. As can be seen, the magnitude of the

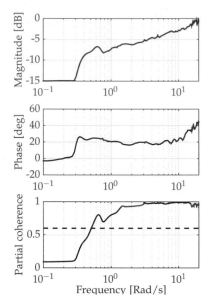

Figure 2.9: Vertical MISO conditioned frequency response a_z/δ_{col} with composite windowing method applied. The dashed line represents a coherence value of 0.6. Values of coherence above this value are considered good.

response increases with the frequency. This phenomenon is caused by the inflow effect, a rotor-body coupling mode that appears in hover condition when the collective position rapidly changes.

Similar rotor-body coupling modes were observed in other on- and off-axis conditioned frequency responses. Therefore, it was concluded that the selected frequency range of excitation for the data collection allowed rotor-body dynamics to be captured. This result is particularly important as it allows the identification of high order models that include rotor/body couplings necessary for implementing reliable control designs.

2.7   Transfer function modeling results

The transfer function modeling was considered as an intermediate
phase of the identification process to acquire information concerning
order of the system, level of input-output couplings and initial parametric values, useful for identifying the final state-space model. In
the transfer function modeling selecting the right model structure is
crucial to avoid over-parameterizations, which generate poor predictive capabilities. Therefore, the model structure selection is usually
done by taking into account the specific application, the frequency
range of interest and the physical meaning associated to the selected
input-output response. In this work, different models were considered
to fit the data response of each input-output axis over the frequency
range of interest $(0.3 - 16 \text{ rad/s})$.

The pitch-rate response q/δ_{lon} and the vertical-acceleration response a_z/δ_{col} will be now considered in detail. These responses are
particularly important as they showed rotor-body coupling modes
that will be taken into account for selecting the transfer functions
orders.

2.7.1   Pitch response

For the identification of the pitch-rate response to the longitudinal
stick input q/δ_{lon}, models with order less than 4 were found not
capable of representing some important dynamic modes, showing cost
functions well above the guideline limit of 100. Therefore, two models
were selected of 4th and 6th order, respectively, associated with the
pitch response physical meaning and able to achieve a good fitting over
the frequency range with high coherence (above 0.6). The bode plots
of the two models is shown in Figure 2.10. The identified 4th order
transfer function model is based on the theory presented in [Heffley,
1979]. This model considers two pairs of complex poles to identify
the low-frequency longitudinal modes of the fuselage dynamics. An
equivalent time delay is added to represent the effective delays caused
by sensor filtering, linkage dynamics between the stick and the rotor,
and additional non-modeled high frequency rotor dynamics. From the
minimization of the cost function of Eq. 2.5, the following transfer

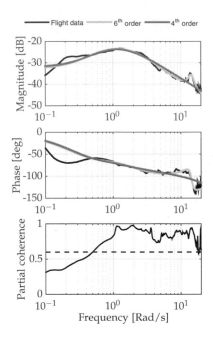

Figure 2.10: Bode plot transfer function models for the pitch response q/δ_{lon}. The dashed line represents a coherence value of 0.6. Values of coherence above this value are considered good.

function model was obtained:

$$\frac{q}{\delta_{lon}} = \frac{0.14(1.317)[-1, 0.361]e^{-0.023s}}{[-1, 0.733][0.947, 1.33]}$$

2.7

where the shorthand notation $[\xi,\omega]$ indicates a pair of complex conjugate roots $s^2+2\xi\omega s+\omega^2$ with damping ratio ξ and natural frequency ω expressed in rad/sec, while $(1/T)$ indicates a single root $s+(1/T)$. The first pair of poles at 0.733 rad/sec models the unstable longitudinal phugoid mode. It can be noticed that the identified damping ratio of this pair is equal to -1. Therefore, the two poles are unstable and real. The second pair at 1.33 rad/sec is associated with the fuselage short period mode. The effect of this mode is a decrease both in magnitude and phase, as can be seen in Figure 2.10. Finally, an equivalent time delay of 0.023 seconds is identified, which accounts for residual high frequency rotor dynamics. The value of the associated cost function is $J = 37.94$, well within the guideline boundary of 100. This result is reflected in Figure 2.10, where it is possible to notice how the 4th order model presents a good level of accuracy over the entire frequency range of interest. However, the model is unable to adequately capture the high frequency lead-lag mode recognized during the nonparametric identification at around 14 rad/sec.

An important measure usually considered to evaluate the fidelity of the identified model with respect to the flight response is the error-response function defined as:

$$\varepsilon_{model}(f) = \frac{T(f)}{\hat{T}_c(f)}$$

2.8

Magnitude and phase of the error-response function associated to the identified 4th order model are shown in Figure 2.11. The mismatch

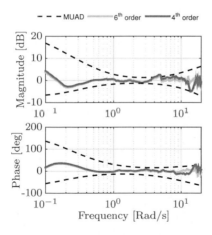

Figure 2.11: Error transfer function of the q/δ_{long} response for the 4th and the 6th order model with Maximum Unnoticeable Added Dynamics (MUAD) boundaries.

boundaries defined in the MIL-STD-179 are also shown and represent the Maximum Unnoticeable Added Dynamics (MUAD) limits, useful for handling qualities studies [Tischler and Remple, 2012]. As long as these boundaries are not exceeded, an experienced helicopter pilot would hardly detect a divergence in the modeled response characteristics [Tischler, 1995]. Figure 2.11 confirms the model incapability to capture the lead-lag mode at 14 rad/sec. Nevertheless, the error value remains within the MUAD mismatch boundaries. Therefore, it can be concluded that the considered 4th order transfer function model is well suited for handling-qualities analyses, but would not be appropriate for control system designs, in which high frequency rotor dynamics are crucial.

A better result was obtained with the 6th order model, able to capture also the regressive lead-lag mode effects as can be seen from the bode plot in Figure 2.10. From the minimization of the cost function of Eq. 2.5, the following transfer function model was obtained:

$$\frac{q}{\delta_{lon}} = \frac{0.11(3.928)[-1, 0.327][0.213, 14.265]e^{-0.019s}}{[-1, 0.683][0.93, 2.065][0.1, 14.336]} \qquad \boxed{2.9}$$

with an associated cost function value $J = 6.17$. In this 6th order model, the phugoid mode is represented with a pair of real unstable poles located at 0.683 rad/sec. The second pair of highly damped complex poles at 2.065 rad/sec models the short period. The rotor lead-lag mode is modeled with a dipole (a pair of conjugate poles at 14.336 rad/sec and a pair of conjugate zeros at 14.265 rad/sec). Finally, a time delay of 0.019 seconds models the residual high frequency rotor dynamics. The error function in Figure 2.11 remains within the boundaries over the entire frequency range of interest, which means that a pilot would consider the model responses almost indistinguishable from the actual flight response. It can be concluded that this model is well suited for both handling qualities evaluation and augmented control system design, due to its ability to capture rotor DOFs relevant for this purpose.

2.7.2 Heave response

The transfer function identification of the vertical response is now considered to investigate the inflow dynamic effects shown during the non-parametric identification phase. Modeling this phenomenon is particularly important to create simulation models in hover that can

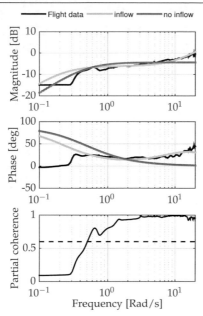

Legend: Flight data — inflow — no inflow

Figure 2.12: Bode plot transfer function models for the vertical-acceleration response due to the collective input a_z/δ_{col}. Two models are compared, one capable of capturing the inflow and the other one unable (no inflow). The dashed line represents a coherence value of 0.6. Values of coherence above this value are considered good.

reproduce the vertical acceleration cues due to rapid collective changes [Ham et al., 1995]. Two transfer function models were considered to fit the vertical-acceleration response due to the collective input a_z/δ_{col}. The first one implements the vertical velocity to collective as a first order function [Ham et al., 1995]. The vertical acceleration response is then obtained by multiplication with the Laplace constant s:

$$\frac{a_z}{\delta_{col}} = \frac{0.6035(0)}{(0.5081)} \qquad \text{2.10}$$

The associated bode plot is shown in Figure 2.12. The related cost function is $J = 145.8$, above the guideline of 100. This is due to the model's failure to reproduce the inflow dynamics at high frequencies.

However, it is interesting to notice that the error function in Figure 2.13 exceeds the MUAD boundaries only at mid-frequencies.

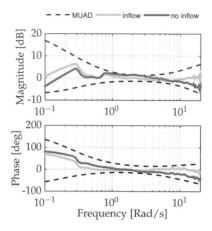

Figure 2.13: Error transfer function of the vertical-acceleration response due to the collective input a_z/δ_{col} with Maximum Unnoticeable Added Dynamics (MUAD) boundaries. Two models are compared, one capable of capturing the inflow and the other one unable (no inflow)

A second model was considered by adding a zero and a delay to represent the inflow dynamics neglected with the first model. The model obtained from the minimization is:

$$\frac{a_z}{\delta_{col}} = \frac{0.0476(0)(10.3384)e^{-0.0284s}}{(0.2364)} \tag{2.11}$$

with an associated cost function $J = 15.98$. The bode plot is shown in Figure 2.12. It can be noticed how the fitting is improved with respect to the first model as the inflow dynamics is now explicitly modeled. The improvement is also reflected in Figure 2.13, where a decrease of the error-response function can be seen at mid- and high-frequencies. From these results it can be concluded that the second model is preferable both for handling qualities evaluations and for control design studies since inflow dynamic effects are adequately identified.

The transfer function modeling was performed for all primary control axes and similar results were obtained as those shown for

the pitch and the heave responses. The information gained from the
transfer function identification phase represented a starting point for
the implementation of the final state-space model considered in the
next sections.

2.8   Extended state-space identification

The state-space identification model is generally obtained from the
helicopter dynamic equations expanded in a Taylor series and trun-
cated at the first order. The unknown coefficients to be identified
are the stability and control derivatives that result from the Taylor
series representation. The model structure used for the state space
identification is the following:

$$M\dot{\mathbf{x}} = F\mathbf{x} + G\mathbf{u} \qquad \text{2.12}$$

$$\mathbf{y} = H_0\mathbf{x} + H_1\dot{\mathbf{x}} \qquad \text{2.13}$$

The selected inputs of the state space model are the four pilot controls

$$\mathbf{u} = [\delta_{lat}, \delta_{lon}, \delta_{ped}, \delta_{col}] \qquad \text{2.14}$$

the outputs are the helicopter response signals measured during the
data collection

$$\mathbf{y} = [u, v, w, p, q, r, a_x, a_y, a_z, \phi, \theta] \qquad \text{2.15}$$

and the state vector is composed of 20 states that represent the
rotor-body dynamics.

$$\mathbf{x} = [u, v, w, p, q, r, \phi, \theta, \beta_{1c}, \beta_{1s}, x_1, x_2, \eta_q,$$
$$y_1, y_2, \eta_p, v, \beta_0, \dot{\beta}_0, \eta_{C_t}] \qquad \text{2.16}$$

The M matrix includes parameters that depend on the derivative of
the state variables, the F matrix includes the stability derivatives
and the G matrix includes the control derivatives. Finally, the two
matrices H_0 and H_1 allow the outputs to be expressed in terms of
state variables and of their derivatives.

As seen in the transfer function modeling, rotor-body coupling
modes (flap, lead-lag and inflow) were captured in the collected data.

To include these modes into the state-space model, a hybrid fully coupled formulation was considered. The equations of this formulation will be now briefly described. The meaning of all symbols used in the equations can be found in the Nomenclature.

The rotor/flap body coupling was modeled by adding two coupled first order equations as presented in [Tischler and Remple, 2012]. The two equations are reported below:

$$\tau_{f_1}\dot{\beta}_{1s} = -\beta_{1s} + Lf_{\beta_{1c}}\beta_{1c} + \tau_{f_1}p + Lf_{\delta_{lon}}\delta_{lon} + Lf_{\delta_{lat}}\delta_{lat} \qquad \boxed{2.17}$$

$$\tau_{f_2}\dot{\beta}_{1c} = -\beta_{1c} + Mf_{\beta_{1s}}\beta_{1s} + \tau_{f_2}q + Mf_{\delta_{lon}}\delta_{lon} + Mf_{\delta_{lat}}\delta_{lat} \qquad \boxed{2.18}$$

where τ_{f_1} and τ_{f_2} are the lateral and the longitudinal flapping time constants, respectively, while β_{1c} and β_{1s} are the states associated with the longitudinal and the lateral flap dynamics, respectively.

Equations (2.17,2.18) are coupled to the roll and the pitch acceleration responses:

$$\dot{p} = L_u u + L_v v + L_w w + L_r r + L_{\beta_{1s}}\beta_{1s} + \\ + L_{\delta_{ped}}\delta_{ped} + L_{\delta_{col}}\delta_{col} \qquad \boxed{2.19}$$

$$\dot{q} = M_u u + M_v v + M_w w + M_r r + M_{\beta_{1c}}\beta_{1c} + \\ + M_{\delta_{ped}}\delta_{ped} + M_{\delta_{col}}\delta_{col} \qquad \boxed{2.20}$$

Here the effects of the state-space control derivatives ($L_{\delta_{lon}}$, $M_{\delta_{lon}}$, $L_{\delta_{lat}}$, $M_{\delta_{lat}}$) and those of the state-space stability derivatives (L_q, M_q, L_p, M_p) are included in the flapping derivatives ($L_{\beta_{1s}}$, $M_{\beta_{1c}}$).

Furthermore, Eqs. (2.17,2.18) are coupled to the longitudinal and the lateral linear acceleration equations by adding the state derivatives $X_{\beta_{1c}}$ and $Y_{\beta_{1s}}$:

$$\dot{u} = -W_0 q + V_0 r + X_u u + X_v v + X_w w + X_r r + \\ + X_{\beta_{1c}}\beta_{1c} + X_{\delta_{ped}}\delta_{ped} + X_{\delta_{col}}\delta_{col} \qquad \boxed{2.21}$$

$$\dot{v} = -U_0 r + W_0 p + Y_u u + Y_v v + Y_w w + Y_p p + Y_r r + \\ + Y_{\beta_{1s}}\beta_{1s} + Y_{\delta_{ped}}\delta_{ped} + Y_{\delta_{col}}\delta_{col} \qquad \boxed{2.22}$$

Including the lead-lag dynamics is generally more difficult as no physical models are available in literature. Usually, the lead-lag is implemented as a second order dipole appended to the angular rate responses [Fletcher, 1995a; Tischler and Cauffman, 1992]. In this way the pitch-rate response with respect to longitudinal stick deflection becomes:

$$
\left(\frac{q}{\delta_{lon}}\right)_{lead-lag} = \left(\frac{q}{\delta_{lon}}\right) \cdot \frac{K_p(s^2 + 2\xi_q\omega_q s + \omega_q{}^2)}{s^2 + 2\xi_{ll_r}\omega_{ll_r}s + \omega_{ll_r}^2}
\tag{2.23}
$$

The same definition holds for the roll-rate response. The initial values of the dipole parameters are usually fixed to the values identified in the transfer function modeling.

In this thesis the dipole form was directly converted into a canonical form and included in the state-space equations as described in [Quiding et al., 2008]. The canonical form for the pitch axis was implemented as follows:

$$
\begin{bmatrix} 1 & 0 & 0 \\ 0 & 1 & 0 \\ 0 & -1 & 1 \end{bmatrix} \cdot \begin{bmatrix} (\dot{x}_1)_q \\ (\dot{x}_2)_q \\ (\dot{\eta})_q \end{bmatrix} = \begin{bmatrix} 0 & 1 & 0 \\ -\omega_{ll_r}{}^2 & -2\xi_{ll_r}\omega_{ll_r} & 0 \\ \omega_q{}^2 & 2\xi_q\omega_q & 0 \end{bmatrix} \cdot \begin{bmatrix} (x_1)_q \\ (x_2)_q \\ (\eta)_q \end{bmatrix} + \begin{bmatrix} 0 \\ K_q \\ 0 \end{bmatrix} [q]
\tag{2.24}
$$

with $\dot{\eta}_q = q_{lead-lag}$. By considering the same state-space canonical form in the roll axis, 9 state variables were introduced: x_1, x_2, η_q for the longitudinal axis and y_1, y_2, η_p for the lateral axis.

Another important dynamic mode captured in the collected data is the coupled fuselage/coning-inflow. This dynamics was implemented through the analytical model proposed by Chen and Hindson [Chen and Hindson, 1986]. In this model the coning/inflow dynamic equations provide three more states: the inflow ν, the coning angle β_0 and the coning rate $\dot{\beta}_0$.

The inflow is modeled as:

$$
\dot{\nu} = \frac{-75\pi\Omega}{32}\left(\bar{\nu}_0 + \frac{a\sigma}{16}\right)C_0\nu + \nu_\beta\dot{\beta}_0 + \\ + \frac{25\pi\Omega^2 R}{32}\left(\frac{a\sigma}{8}\right)C_0 K_{\theta_0}\delta_{col}
\tag{2.25}
$$

The coning dynamics is instead expressed as a second order differential equation:

$$
\ddot{\beta}_0 = \frac{-\Omega\gamma}{8} - \Omega^2\beta_0 - \frac{\Omega\gamma}{6R}\nu + \frac{\Omega^2\gamma}{8}K_{\theta_0}\delta_{col}
\tag{2.26}
$$

The coning/inflow is coupled to the fuselage with a perturbation thrust coefficient introduced as a fictitious state derivative $\dot{\eta}_{C_t}$:

$$C_0 \dot{\eta}_{C_T} = \frac{0.543}{\Omega^2 R}\dot{\nu} + \frac{4\bar{\nu}_0}{\Omega R}C_0 \nu + \frac{4\bar{\nu}_0}{3\Omega}C_0 \dot{\beta}_0 \qquad \boxed{2.27}$$

Finally, the vertical acceleration equation is expressed as:

$$\dot{w} + \left[\frac{\rho \pi R^2 (\Omega R)^2}{m}\right]\dot{\eta}_{C_T} = Z_u u + Z_v v + Z_w w + Z_p p +$$
$$+ Z_q q + Z_r r + Z_{\delta_{lon}}\delta_{lon} + \qquad \boxed{2.28}$$
$$+ Z_{\delta_{lat}}\delta_{lat} + Z_{\delta_{ped}}\delta_{ped}$$

A final important aspect to consider is the knowledge of the center of gravity position. Generally, accelerations are not directly measured at the center of gravity. Therefore, a correction needs to be done. Although measurements of longitudinal (x_a) and lateral (y_a) center of gravity positions can be done through specific softwares provided by helicopter companies, the vertical (z_a) position is generally hard to measure. For this reason, z_a was left as unknown and identified through the equations of longitudinal and lateral accelerations:

$$a_x = \dot{u} + z_a \dot{q} - y_a \dot{r}$$
$$a_y = \dot{v} - z_a \dot{p} + x_a \dot{r} \qquad \boxed{2.29}$$

The resulting hybrid state space model implemented through the equations presented in this section contains 12 DOFs that account for rotor/flap body coupling, lead-lag and coning-inflow dynamics. All equation presented here contain many unknown parameters which are to be identified. However, some parameters are generally fixed, based on measurements and values provided in literature.

Starting from this model, the procedure described in [Tischler and Remple, 2012] was implemented to identify the unknown parameters. In this procedure, a parameter can be eliminated from the model if considered unidentifiable or without any physical meaning. This is done based on the insensitivities and the Cramer-Rao bounds associated with the model's parameters. The insensitivity is a measure of the change in the cost function value (Eq. 2.6) due to a change in a model's parameter. The Cramer-Rao bound is an estimate of the standard deviation associated with an identified parameter. It gives an indication of the identified value reliability and of its correlation with other parameters. By discarding all parameters considered insignificant or with high correlation, it is possible to obtain a reduced model structure with physical meaning while avoiding over-parameterization. In the next section the results obtained by applying this procedure are presented.

2.9   State space modeling results

The unknown parameters of the hybrid model were identified by matching 44 frequency responses computed in the non-parametric identification. For each response 20 frequency points were selected in the frequency range of interest (0.3-16 rad/sec) with relative coherence function larger than 0.5, based on the guidelines presented in [Tischler and Remple, 2012].

As no direct rotor state measurements were collected, parametric correlation problems were reduced by considering constraints that allow physically meaningful parametric values to be obtained [Tischler and Cauffman, 1992]. The lateral and the longitudinal flapping time delays τ_{f_1} and τ_{f_2} (Eq. 2.17, 2.18) were set to be the same ($\tau_f = \tau_{f_1} = \tau_{f_2}$), consistently with the theory predictions [Tischler and Remple, 2012]. Furthermore, their values were fixed to the time delay identified during the transfer function modeling of the roll-rate response (p/δ_{lat}). The initial value of the flapping roll spring $L_{\beta_{1s}}$ (Eq. 2.19) was set to the squared frequency of the roll/flapping mode, obtained in the transfer function identification of the roll-rate response (p/δ_{lat}). Furthermore, the flapping pitch spring $M_{\beta_{1c}}$ initial value (Eq. 2.20) was set to one-third of the roll spring ($L_{\beta_{1s}}$) value, as predicted by theory. Usually, the coupling terms $Lf_{\beta_{1c}}$, $Lf_{\delta_{lon}}$ (Eq. 2.17) and $Mf_{\beta_{1s}}$, $Mf_{\delta_{lat}}$ (Eq. 2.18) are neglected in literature. However, during the identification process a better fitting was achieved by retaining $Lf_{\beta_{1c}}$, $Lf_{\delta_{lon}}$ in the model. The values of $X_{\beta_{1c}}$ and $Y_{\beta_{1s}}$ in Eqs. 2.21 and 2.22 were fixed to g and $-g$, respectively. This choice was based on the assumption that the flapping contribution in the longitudinal force X and the lateral one Y is mainly caused by the tilt of the main rotor thrust vector [Fletcher, 1995a].

Concerning the lead-lag dynamics (Eq. 2.23) the denominator terms (damping ratios ξ_{ll_r} and natural frequencies ω_{ll_r}) were constrained to be the same for both longitudinal and lateral axes, based on symmetry. Moreover, their values were fixed to the ones obtained in the transfer function identification of the pitch-rate response (Eq. 2.9). Finally, the time delays were fixed to the values identified during the transfer function identification.

The state-space identification was possible for 22 of the 44 frequency responses. Most of the secondary responses in the vertical and in the yaw control axes were discarded due to low coherence (less than 0.6) throughout the entire frequency range of interest. The algorithm used to minimize the MIMO cost function (Eq. 2.6) was sensitive to initial parametric values and bounds. As a result of this, different local minima were obtained. An iterative pattern search algorithm is generally used to overcome this issue, as proposed by Tischler [Tischler and Remple, 2012]. However, this procedure did not achieve satisfactory results as poor fits were obtained in many responses (high values of cost-function). Therefore, a different procedure was considered consisting of selecting a larger number of frequency points (50 instead of 20). The points were included if the associated partial coherence function ($\hat{\gamma}^2$) was larger than 0.3 (instead of the 0.5 usually adopted in literature). Although this approach was not based on theoretical analyses, it was assumed that including additional points with lower coherence would increase the frequency content of the data used for the model fitting. In the optimization, the selected frequency points were weighted with the associated coherence values. By using this procedure, solutions were found less sensitive to initial parametric values and boundaries.

However, low coherence values are generally associated with process noise, nonlinearities, lack of input excitation or lack of rotorcraft response. For this reason, the original guidelines (20 frequency points and partial coherence $\hat{\gamma}^2 > 0.5$) were again applied to fit the final state-space model. This time though, local minima were avoided by selecting initial parametric values and boundaries based on the solutions obtained from the procedure described above.

The identified parameters of the final model are listed in Table 2.1 with associated Cramer-Rao bounds and insensitivities. The model is characterized by an acceptable theoretical accuracy since only one parameter has relative Cramer-Rao bound larger than 20% and insensitivity value larger than 10%, which are usually defined in literature as desired maximum values. During the identification it was verified that dropping this parameter from the model led to a large marginal increase in the overall model cost function (Eq. 2.6). Therefore, the model was not further reduced.

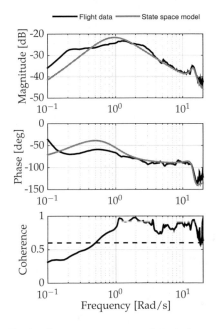

Figure 2.14: Bode plot state space model pitch response to longitudinal cyclic q/δ_{lon}.

The pitch-rate and the heave responses are now considered in details and compared to the results obtained in the transfer function identification. Figure 2.14 shows the pitch-rate response to the longitudinal stick input (q/δ_{lon}). As expected, a less accurate fitting is achieved at low frequencies with respect to the transfer function identification (Figure 2.10) since 22 responses are simultaneously fitted in the state-space identification. Nevertheless, the obtained pitch-rate cost function is $J_{q/\delta_{lon}} = 74.38$, well below the guideline limit of 200. The final model is able to represent the low-frequency phugoid oscillation with a pair of unstable complex poles located at 0.541 rad/sec. This frequency value is very close to the one identified in the transfer function modeling (0.683 rad/sec). As can be noticed in Figure 2.14, the model is able to represent the lead-lag mode. The identified parameters associated with this mode are also listed in Table 2.1. Of these parameters, the terms corresponding to the lead-lag dipole numerator (ξ_{ll_r},ω_{ll_r}) were left to be determined during the identification. Those corresponding to the denominator (ξ_q,ω_q) were fixed to the values identified during the transfer function modeling (Eq. 2.9).

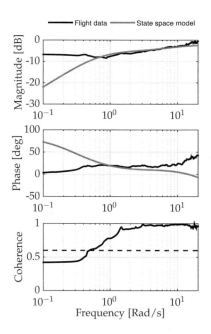

Figure 2.15: Bode plot state space models vertical response to collective a_z/δ_{col}.

The vertical acceleration response to the collective input (a_z/δ_{col}) is shown in Figure 2.15. The good quality of the fit in the frequency range of interest $(0.3 - 16$ rad/s$)$ is reflected in the low cost function $J_{a_z/\delta_{col}} = 62.98$. It is interesting to notice that despite the inclusion of the inflow dynamics and the high value of the associated partial coherence function, the model is not able to reproduce the phase increase at high frequencies. This was attributed to the representation of the inflow dynamics in state-space form, as the representation used for the transfer function modeling was able to reproduce this response behavior (Figure 2.12).

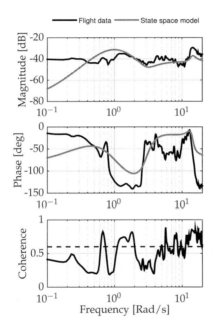

Figure 2.16: Bode plot state space models roll response to longitudinal cyclic p/δ_{lon}.

In the state-space identification the p/δ_{lon} frequency response (Figure 2.16) achieved the worst fitting with a cost function of 315. This result was attributed to the particularly high random error of the response at high frequencies. However, the response was not discarded since it showed a high partial coherence function $(\hat{\gamma}^2)$ in a sufficient number of points within the frequency range of interest. Furthermore, despite the high associated cost function, the identified state-space model was able to resemble the measured response behavior in the frequency range of interest. In the next section, the identified state-space model will be validated in the time domain by using the data collected during the doublet-maneuvers performed in the flight tests.

For the frequency responses that are not shown here, the identification provided similar results as the ones presented in this section. The primary responses are shown in Appendix C.

Table 2.1: Identified stability and control derivatives

Derivative	Value	C.R. (%)	Insens. (%)	Derivative	Value	C.R. (%)	Insens. (%)
X_u	-0.354	12.757	5.396	X_{lat}	0^a	n/a	n/a
X_v	0^a	n/a	n/a	X_{lon}	0^a	n/a	n/a
X_w	0^a	n/a	n/a	X_{ped}	0^a	n/a	n/a
X_p	0^a	n/a	n/a	X_{col}	0^a	n/a	n/a
X_q	0^a	n/a	n/a	Y_{lat}	0^a	n/a	n/a
X_r	0^a	n/a	n/a	Y_{lon}	0^a	n/a	n/a
$X_{\beta_{1c}}$	9.810^b	n/a	n/a	Y_{ped}	0^a	n/a	n/a
Y_u	0^a	n/a	n/a	Y_{col}	0^a	n/a	n/a
Y_v	-0.144	35.945	16.241	Z_{lat}	0^a	n/a	n/a
Y_w	0^a	n/a	n/a	Z_{lon}	0^a	n/a	n/a
Y_p	0^a	n/a	n/a	Z_{ped}	0^a	n/a	n/a
Y_q	0^a	n/a	n/a	Z_{col}	0^a	n/a	n/a
Y_r	0^a	n/a	n/a	L_{lat}	0^a	n/a	n/a
$Y_{\beta_{1s}}$	-9.810^b	n/a	n/a	L_{lon}	0^a	n/a	n/a
Z_u	0^a	n/a	n/a	L_{ped}	0^a	n/a	n/a
Z_v	0^a	n/a	n/a	L_{col}	0^a	n/a	n/a
Z_w	-0.324	15.328	6.944	M_{lat}	0^a	n/a	n/a
Z_p	0.754	16.003	7.318	M_{lon}	0^a	n/a	n/a
Z_q	0^a	n/a	n/a	M_{ped}	0^a	n/a	n/a
Z_r	0^a	n/a	n/a	M_{col}	0^a	n/a	n/a
L_u	0^a	n/a	n/a	N_{lat}	0^a	n/a	n/a
L_v	0.064	12.235	3.461	N_{lon}	0^a	n/a	n/a
L_w	0^a	n/a	n/a	N_{ped}	0.312	2.434	1.089
L_r	0^a	n/a	n/a	N_{col}	0^a	n/a	n/a
$L_{\beta_{1s}}$	44.022	7.090	0.878	$Lf_{\delta_{lon}}$	0.0074	9.913	0.158
M_u	-0.042	9.092	2.836	$Lf_{\delta_{lat}}$	0.0067	7.139	0.805
M_v	0^a	n/a	n/a	$Mf_{\delta_{lon}}$	0.0095	5.556	0.164
M_w	0^a	n/a	n/a	$Mf_{\delta_{lat}}$	0^a	n/a	n/a
M_p	-0.304	8.988	3.472	K_{θ_0}	-0.0055	3.083	1.292
M_r	0^a	n/a	n/a	τ_{lat}	0.001^b	n/a	n/a
$M_{\beta_{1c}}$	9.053	5.195	0.905	τ_{lon}	0.019^b	n/a	n/a
N_u	0^a	n/a	n/a	τ_{ped}	0^b	n/a	n/a
N_v	0^a	n/a	n/a	τ_{col}	0.028^b	n/a	n/a
N_w	-0.254	7.496	3.106				
N_p	-0.288	19.960	8.839				
N_q	0^a	n/a	n/a				
N_r	-1.193	6.153	2.452				
$N_{\beta_{1s}}$	7.106	10.185	2.548				
τ_f	-0.026^b	n/a	n/a				
$Lf_{\beta_{1c}}$	-0.745	8.081	0.169				
$Mf_{\beta_{1s}}$	0^a	n/a	n/a				
ν_β	-18.962	n/a	n/a				
C_0	0.639	n/a	n/a				
K_p	1.062	7.091	1.649				
ω_p^2	187.886	4.749	1.259				
$2\xi_p\omega_p$	4.869	12.289	4.891				
$\omega_{ll_r}^2$	-205.521^b	n/a	n/a				
$2\xi_{ll_r}\omega_{ll_r}$	-2.867^b	n/a	n/a				
K_q	1.205	9.699	1.965				
ω_q^2	215.379	7.370	1.697				
$2\xi_q\omega_q$	6.689	16.322	6.383				
$\omega_{ll_{r_2}}^2$	-205.521^b	n/a	n/a				
$2\xi_{ll_{r_2}}\omega_{ll_{r_2}}$	-2.867^b	n/a	n/a				
z_a	0.143	19.415	5.871				

[a] Eliminated during model determination

[b] Fixed value in model

2.10   Validation of the state-space model

The identified state-space model was validated in the time domain to test its predictive capabilities with respect to different inputs (doublets) than the ones considered during the identification (frequency sweeps). Figures 2.17 and 2.18 show the difference between the flight test data and the model responses to a doublet given in one of the four control axes.

As can be seen in Figure 2.17a, a very good agreement is achieved in the lateral axis between the model and the collected data. However, the model exaggerates the yaw-rate response (r/δ_{lat}) at low frequencies. A satisfactory results is also achieved in the longitudinal axis (Figure 2.17b). In particular, it can be noticed that the off-axis roll-rate response (p/δ_{lon}) follows the data well, despite the poor fit in the associated frequency response (Figure 2.16). As for the lateral axis, the yaw-rate response (r/δ_{lon}) shows some discrepancies.

The yaw and the vertical axes have excellent agreements in the on-axes responses but the model predictive capabilities worsen in the off-axes responses (Figures 2.18a and 2.18b). This reflects the result obtained during the identification, in which most of the off-axes responses were discarded because of low frequency content over the entire frequency range of interest. An important limitation of the model is its inability to properly predict the helicopter yaw/heave coupling (r/δ_{col}). This was attributed to a high coupling between collective and pedals inputs given by the pilot during the flight tests, which could also explain the discrepancies in the yaw-rate responses observed in both longitudinal and lateral axes. Despite these limitations, the model shows good predictive capabilities and the ability to capture important characteristics for handling qualities analysis and control system design.

A final model assessment was done by performing hover and low speed closed-loop control task maneuvers with a helicopter pilot in the CyberMotion Simulator [Nieuwenhuizen and Bülthoff, 2013]. The pilot assessed the model in visual-only-condition first and in visual-plus-motion-condition after. The simulator cabin was equipped with a pilot seat, a conventional center-stick cyclic, a collective lever and rudder pedals. A virtual environment representing an airport

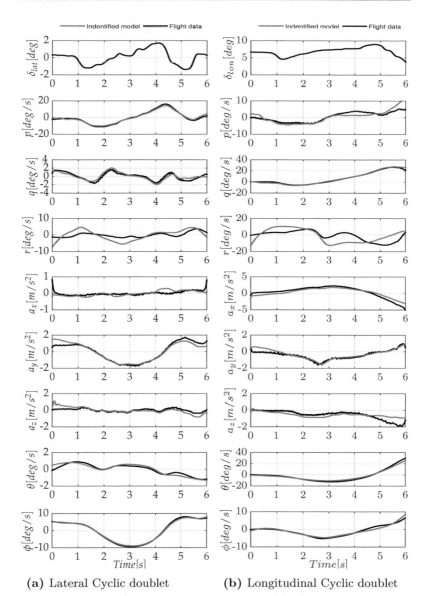

(a) Lateral Cyclic doublet **(b)** Longitudinal Cyclic doublet

Figure 2.17: Time validation with doublets

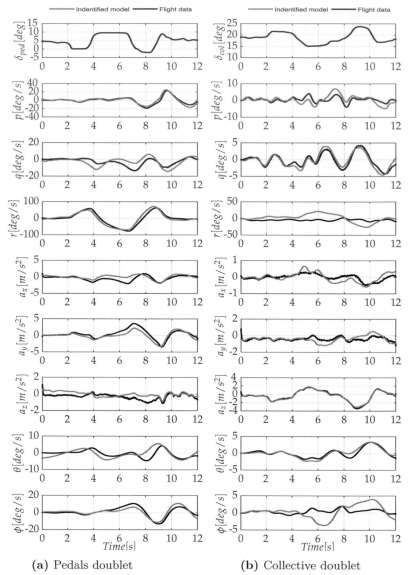

(a) Pedals doublet   (b) Collective doublet

Figure 2.18: Time validation with doublets

was projected inside the simulator cabin showing a helicopter pilot perspective.

All trials started in hover with all controls at zero position. The pilot had full control of all axes. First, longitudinal and lateral doublets were performed. The pilot appreciated the coupling between the two axes that he described as "quite realistic". The motion in both axes was considered "very good". The vertical and yaw axes were analyzed next. Again some doublets were performed in both axes. The pilot described the pedals as "more sensitive than expected". However, he appreciated the range of motion associated with the yaw axis. The yaw-heave coupling was considered "different than in a real helicopter" in agreement with the validation results. However, he indicated to be "positively surprised by the model's capability of capturing these helicopter characteristics". Finally, the pilot was instructed to perform maneuvers in hover and low speed flight regime. The model was overall assessed as "quite good and realistic". The positive feedback given by the helicopter pilot confirmed the analysis obtained from the time validation.

2.11 Conclusions

This chapter presented results on the implementation of a fully coupled state-space model of a Robinson R-44 civil light helicopter in hover. The data collection phase was described in details with particular attention devoted to the development of the measurement setup. From the obtained frequency responses, it was deduced that the chosen frequency range of excitation allowed important rotor dynamic modes to be captured. The pitch-rate response with respect to the longitudinal cyclic and the heave response to the collective were considered in details throughout the chapter.

First, a transfer function identification was performed and important information was gained concerning order of the system, level of input-output couplings and initial values for some state-space parameters. Then, the main dynamic equations necessary to represent rotor-body couplings in state-space form were introduced and the state-space identification was performed. A new procedure was

proposed to limit the sensitivity of the cost function minimization
algorithm to initial parametric values and bounds. The final model
showed good predictive capabilities, being able to capture high frequency rotor-body coupling dynamics. Therefore, the model was
assessed well suited for handling qualities and control system design
studies, two main goals of this thesis.

This chapter proposed a procedure to adapt system identification
methods for applications in the civil helicopter field. It showed that
frequency domain identification methods are mature enough to be
implemented in numerical computing environments like MATLAB®.
Furthermore, it showed that the civil field could benefit from the use
of system identification. Significant information could be gained for
handling qualities studies and augmented control system designs by
applying a lean and tested procedure for data collection as the one
proposed here.

The identification results showed that the pilot training before the
data collection represents a crucial step. Furthermore, checking that
all control axes are adequately excited while performing frequency
sweeps and doublets is important to ensure the collection of reliable
data.

The results presented in this chapter complete the first objective
of the thesis and represent a starting point for the next objectives
presented in the following sections.

CHAPTER 3

Robust control methods to augment an identified helicopter model

This chapter presents the implementation of classical robust control stategies considered to augment the identified state-space helicopter model described in Chapter 2. Aim of this study is to enhance stability and controllability of the helicopter model to achieve response types and Handling Qualities of a new category of aircrafts, the Personal Aerial Vehicles. Two control methods are considered to develop the augmented control systems, H∞ and μ-synthesis. The resulting augmented systems are compared in terms of robust stability, nominal performance and robust performance. Furthermore, differences, advantages and limitations of the implemented control architectures are highlighted with respect to the selected reference dynamics.

The contents of this chapter are based on:

Paper title   Augmented Systems for a Personal Aerial Vehi-
                     cle Using a Civil Light Helicopter Model

Authors   S. Geluardi, F. M. Nieuwenhuizen, L. Pollini
                     and H. H. Bülthoff

Published in   Proceedings of the American Helicopter Society
                     71th Annual Forum, Virginia Beach, USA, 2015

H ELICOPTERS are characterized by complex dynamics that make them difficult to model and control. In the last few decades a big effort has been made to enhance helicopters stability and controllability and to meet demanding specifications like the Handling Qualities (HQs) requirements of the Aeronautical Design Standard-33 (ADS-33E-PRF) [Postlethwaite et al., 2005; Theodore et al., 2014]. Reasons for motivating this kind of research are the possibility to decrease pilot workload and improve safety. Furthermore, helicopter models obtained through system identification and highly complex nonlinear models have allowed the implementation of control augmented system in simulation. As a result of this, cost and safety-risk reduction have been obtained by avoiding controllers tuning and validation through extensive flight tests [Theodore et al., 2014; Walker and Postlethwaite, 1996].

So far, most of the research on helicopter control has been conducted in the military field. Therefore, studies to improve helicopter HQs have been focusing on experienced pilots. Both scientific and civil helicopter communities have shown no interest in applying control augmented approaches to achieve helicopter dynamics considered intuitive and easy to control for novice pilots.

However, with the definition of a new kind of vehicles, the Personal Aerial Vehicles, the myCopter project [Bülthoff et al., 2011] extended the concept of HQs to non-expert pilots. In particular, this research showed that classical helicopter response types such as attitude command attitude hold (ACAH) or rate command attitude hold (RCAH) are not suitable for non-expert pilots, whereas translational rate command (TRC) response types allow non-expert pilots to perform complex maneuvers with adequate performance. These results have inspired a new branch of study with the aim to investigate whether augmentation strategies can be applied to actual vehicle dynamics to achieve HQs defined for PAVs. In this thesis, civil light helicopters are proposed as possible PAVs candidates. In particular, the vehicle dynamics considered here is that of an identified Robinson R44 Raven II helicopter in hover (Chapter 2).

The goal of this chapter is the implementation of robust control techniques that can augment the helicopter dynamics to achieve response types and HQs defined for PAVs. Two control methods are considered here, the H∞ and μ-synthesis. These robust control designs were chosen because of the many studies documented in literature in which they are applied on helicopter models [Postlethwaite et al., 2005]. In particular, the H∞ control method was selected for its capability to deal with Multi-Input Multi-Output (MIMO) linear systems subject to uncertainties and external disturbances [Walker et al., 2000]. Furthermore, its frequency based approach is well suited for models that include high frequency content, like the identified helicopter model considered in this thesis. Another advantage is the possibility to easily integrate the HQs specifications proposed for PAVs as they are based on frequency domain criteria. The limitation of the H∞ control system approach is that performance requirements cannot be ensured against model parametric uncertainties, thus robust performance is not guaranteed. For this reason, the μ-synthesis robust control technique was also considered in this thesis and compared with the H∞ approach. The advantage of the μ-synthesis is the possibility to include the model uncertainties into the optimization problem to achieve robustness [Zhou and Doyle, 1999]. Moreover, the μ-synthesis is a frequency based method as well as the H∞ one, thus presents all advantages shown before. The only disadvantage is that resulting controllers have typically high orders.

In this chapter H∞ and μ-synthesis robust control techniques are compared against stability and performance requirements. Robustness achieved by the two control techniques are tested against external disturbances and model parametric uncertainties. External disturbances are considered as real atmospheric turbulences that might be experienced in hover and low speed flight regime. Model parametric uncertainties are associated with the Cramer-Rao (CR) bounds computed during the model identification (Chapter 2). A model following approach is implemented to achieve response types and HQs defined for PAVs.

The chapter is organized as follows. First, features of the identified helicopter model that are important for control design purposes are highlighted. Then, a description is given of the design methods used for implementing the H∞ and the μ-synthesis control techniques. The obtained results are described in terms of robust stability, nominal performance and robust performance. Finally, a discussion is given to

highlight differences, advantages and limitations of the implemented
control architectures with respect to the considered requirements.

3.1   Helicopter model

A brief summary is now given of the method presented in Chapter 2 to
identified the helicopter model studied in this thesis. The model was
obtained by performing two hours of flight tests with a Robinson R44
Raven II helicopter in hover trim condition. Two Global Positioning
System antennas and an Inertial Measurement Unit were used to
collect the signals defined as the outputs of the model to identify
(inertial position $[x, y, z]$, attitude $[\phi, \theta, \psi]$, angular rates $[p, q, r]$ and
linear accelerations $[a_x, a_y, a_z]$). Four optical sensors were used to
measure the piloted control input signals (cyclic stick, pedals deflec-
tions and collective lever $[\delta_{lon}, \delta_{lat}, \delta_{ped}, \delta_{col}]$). Data were collected
while performing several piloted frequency sweeps and doublets [Gelu-
ardi et al., 2013]. Then, a frequency domain identification control
method was implemented to obtain a linear model of the helicopter
in hover.

The first step of the identification was a non-parametric analysis
done by computing the conditioned frequency responses (to account
for input-output couplings and input correlations) and by applying
the composite windowing method (to reduce the random error in the
spectral estimates) as described by Tischler [Tischler and Remple,
2012]. Then, a transfer function model was implemented and validated
[Geluardi et al., 2014] to gain information about fundamental dynamic
characteristics, order of the system, level of couplings and to obtain
initial reasonable parametric values useful to build the final state-space
model.

The final step of the system identification process was the implementation of a fully coupled 12 degree-of-freedom state space-model.
An important feature of this model is its capability of accounting for
rotor/flap body coupling dynamics, lead-lag dynamics and coning-
inflow dynamics. These dynamics are very important as they can
limit control feedbacks gains on attitude and rate [Chen and Hindson,
1986; Greiser and Lantzsch, 2013]. Therefore, the identified model is

well suited for implementing reliable control augmentation systems and will be used in this chapter.

3.2   Robust control theory

Control system designs have to take into account all important physical characteristics of a model to ensure good results, especially when applied to complex dynamics like helicopters. Usually, helicopter models are unstable, non-minimum phase and show important rotor-body coupling dynamics. The model identified in this thesis presents all these characteristics, which had to be considered during the control design. A brief theoretical description of the H∞ and the μ-synthesis control methods will now be given. Then, the control design will be presented in details.

3.2.1   H∞ control

The H∞ optimization approach consists of synthesizing a stabilizing controller that minimizes the H∞ norm of the closed-loop response from the exogenous inputs (e.g. commands and disturbances) to the controlled outputs (e.g. tracking errors, actuator signals and performance variables). Generally, sensitivity and co-sensitivity functions are considered for this purpose. In particular, the weighted sensitivity function is minimized to ensure good tracking and disturbance attenuation, while the weighted co-sensitivity functions is minimized to obtain a lower control energy action. The H∞ robust approach solves a sub-optimal problem to guarantee that the infinity norms of the weighted sensitivity and co-sensitivity function are sufficiently small. This control approach allows for stability and performance to be satisfied against external disturbances and noise. However, requirements cannot be ensured with respect to model parametric uncertainties. A full description of the theory of the H∞ approach can be found in literature [Skogestad and Postlethwaite, 2007].

3.2.2 μ-synthesis control

The μ-synthesis control technique allows for stability and performance
to be achieved against model uncertainties. This is in general obtained by transforming all uncertainties into structured ones and by
including them into an optimization problem. This control technique
is based on the structured singular value defined in literature [Zhou
and Doyle, 1999]. The aim of this approach is to find a controller that
ensures a structured singular value less than one, when applied to the
connection of a system transfer function matrix with the associated
normalized uncertainty block. Fulfillment of this condition guarantees
robust performance of the closed-loop system [Boyd and Barratt,
1991]. A disadvantage of this approach is that the resulting controllers have typically high orders, whose reduction while maintaining
robustness and required performance is not always a trivial task. Low
order linear controllers are usually preferable since they are easier
to implement, have higher reliability and are computationally less
demanding. On the contrary, high order controllers lead to high costs,
difficult commissioning, poor reliability and problems in maintenance
[Zhou and Doyle, 1999]. For this reasons, a controller order reduction
will also be implemented here.

3.3   Control design

Both H∞ and μ-synthesis control methods consist of solving an
optimization problem that has to minimize

- for the H∞ control technique: the H∞ norm of the closed-loop
  response from the exogenous inputs to the controlled outputs;

- for the μ-synthesis control technique: the structured singular
  value.

Two main steps are necessary to solve these problems. In the first
step a control scheme is designed and weighting functions are selected
to weight the quantities that have to be minimized. In the second
step a controller is implemented that has to stabilize the system
and minimize the weighted quantities. This two steps will be now
considered for both control designs.

3.3.1 Control System Architecture

The system architecture in Figure 3.1 was considered for both control methods.

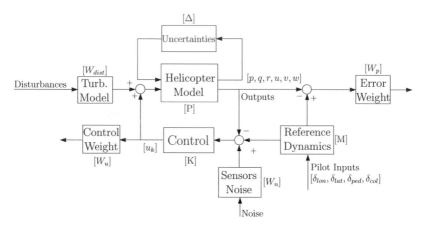

Figure 3.1: Control system architecture

In this control architecture the external inputs are the *Pilot Inputs* $[\delta_{lon}, \delta_{lat}, \delta_{ped}, \delta_{col}]$, the external *Disturbances* $[W_{dist}]$ and the sensors *Noise* $[W_n]$.

The *Helicopter Model [P]* block is the identified linear state-space model in hover trim condition.

For both control methods, robustness specifications of stability and performance need to be tested against model uncertainties, disturbances and noise acting on the system. The *Uncertainties* $[\Delta]$ block contains the parametric uncertainties of the helicopter model associated with the Cramer-Rao (CR) bounds of the identified model parameters. Uncertainties are due to discrepancies between the identified model and the actual helicopter dynamics. The CR bounds were calculated in Chapter 2 according to the theory presented by Tischler [Tischler and Remple, 2012] and represent the expected standard deviation of the identified parameters. Therefore, $\pm 3CR_i$ constitutes for each parameter 99% confidence interval. Changing the value of each parameter within these intervals determines system variations that lead to different responses, thus different HQs characteristics. These variations will be used to evaluate how robust the implemented control systems are in terms of stability and performance.

The *Disturbances* $[W_{dist}]$ are considered here as real atmospheric turbulences that might be experienced in hover and low speed flight regime. No data were collected to identify a specific turbulence model for the R44 helicopter. Therefore, a turbulence model was selected, developed by the German Aerospace Center DLR for the EC135 helicopter. This model is valid in hover and low speed flight conditions. It was obtained by using the Control Equivalent Turbulence Input (CETI) method [Seher-Weiss and von Gruenhagen, 2012] and is composed of four transfer functions driven by a white noise (*Disturbances* in Figure 3.1). These transfer functions are attached to the pitch, roll, yaw and heave control axes and are reported in Eq. 3.1.

$$\frac{\delta_{lon,gust}}{Dist.} = \frac{2.71}{s + 1.57}$$

$$\frac{\delta_{lat,gust}}{Dist.} = \frac{2.56}{s + 1.57}$$

$$\frac{\delta_{ped,gust}}{Dist.} = \frac{7.59}{s + 2.85}$$

$$\frac{\delta_{col,gust}}{Dist.} = \frac{0.473(s + 31.4)}{(s + 0.9891)(s + 7.85)}$$

3.1

The *Sensors Noise* $[W_n]$ block was implemented by considering noise shaping functions based on realistic accelerometer and rate-gyros:

$$W_{acc.} = 0.0002\frac{0.12s + 1}{0.001s + 1}$$

3.2

$$W_{gyros} = 0.06\frac{0.18s + 1}{0.002s + 1}$$

3.3

In order to achieve response types and HQs associated to the PAV dynamics, a model following approach was implemented. The *Reference Dynamics* [M] are Translational Rate Command (TRC) response types for lateral, longitudinal and vertical translational degrees of freedom. In the yaw axis a Rate Command Attitude Hold (RCAH) response type is considered that ensures Level 1 HQs, which corresponds to desirable flying qualities according to the ADS-33-PRF definition [Baskett, 2000]. The transfer functions associated to these responses are:

$$M_{lon} = \frac{u_{ref}}{\delta_{lon}} = \frac{1.05}{1.25s + 1}$$

$$M_{lat} = \frac{v_{ref}}{\delta_{lat}} = \frac{1.05}{1.25s + 1}$$

$$M_{yaw} = \frac{r_{ref}}{\delta_{ped}} = \frac{0.03e^{-0.008}}{0.25s + 1}$$

$$M_{vert} = \frac{w_{ref}}{\delta_{col}} = \frac{e^{-0.20}}{5s + 1}$$

(3.4)

Two more responses were considered (Eq. 3.5) to achieve pitch and roll dynamics defined for PAVs. These responses are Attitude Command Attitude Hold (ACAH) types with associated Level 1 HQs, corresponding to desirable flying qualities according to the ADS-33-PRF definition.

$$M_{pitch} = \frac{\theta_{ref}}{\delta_{lon}} = \frac{0.36e^{-0.008}}{s^2 + 4.22s + 5.49}$$

$$M_{roll} = \frac{\phi_{ref}}{\delta_{lat}} = \frac{0.73e^{-0.008}}{s^2 + 3.51s + 5.49}$$

(3.5)

The control problem defined with this architecture consists of finding a controller [K] that minimizes the control action $[u_k]$ and the difference between the reference dynamics and the helicopter outputs $[u, v, w, p, q, r]$, while ensuring robust stability against uncertainties, external disturbances and sensor noise. The 6 inputs of the controller [K] are the differences between the *Reference Dynamics* and the helicopter *Outputs*. The 4 outputs are the control actions $[u_k]$ applied to the helicopter model.

The selection of the weighting functions ($[W_u, W_p]$) is described in the next subsection.

3.3.2 Weighting Function Selection

The weighting functions selection ($[W_u, W_p]$) is usually a crucial step, in which different aspects have to be taken into account [Bibel and Malycvac, 1992]. Generally, minimizing control actions and difference between system outputs and references is not possible over the entire frequency range because of design constraints and limitations. For example properties of a physical system might limit the frequency range over which feedback gains can be large [Zhou and Doyle, 1999].

An important aspect is the frequency range of interest. Here, a frequency range of [0.3-16] rad/s was selected because of the validity of the identified model within this range. Two other aspects are the control effort and the bandwidth linked to the actuators performance. These two aspects need to be considered to prevent actuators from achieving unfeasible velocities.

Based on these considerations, the weighting functions zeros and poles were fixed as reported in Eqs. 3.7 and 3.8. It is important to notice that the weighting functions were chosen as high pass filters for the control function $[W_u]$ and low-pass filters for the performance function $[W_p]$, as generally considered in literature [Bibel and Malyevac, 1992].

Usually, selecting optimal gains in the weighting functions can be done through experience, several trials and an accurate inspection of the model characteristics. However, applying these classical procedures did not provide controllers able to achieve satisfactory performance results. For example, increasing the control action by limiting the gains $[W_u]$ was not effective because large control gains on the longitudinal and lateral axes led to instability. This effect was attributed to the rotor body couplings included in the identified helicopter model. It is well documented in literature how lead-lag modes can limit the gains on attitude-rate control feedbacks [Greiser and Lantzsch, 2013; Tischler, 1991].

Another difficulty in the weighting functions selection was liked to the tracking of the reference dynamics (equations 3.4, 3.5). In a helicopter the translational dynamics are strictly coupled to the rotational ones. Therefore, achieving good tracking of both translational and rotational PAV rate responses was not a trivial task.

To overcome these issues and select the optimal weighting functions gains, an optimization problem was implemented. In this problem, three objective functions were defined as follows.

1. The cost γ for which the infinity norm of the nominal closed loop system architecture (CL) in Figure 3.1 satisfies the following $||CL||_\infty < \gamma$.

2. The sum of the differences between the translational velocities frequency responses of the augmented system and those of the PAV reference model. Each difference was computed in terms of magnitude and phase as follows:

$$J_i = \sum_{\omega_1}^{\omega_{n_\omega}} \left[\left(|T_{aug_i}| - |T_{ref_i}| \right)^2 + \left(\angle T_{aug_i} - \angle T_{ref_i} \right)^2 \right] \qquad \boxed{3.6}$$

with T_{aug_i} representing a SISO response of the augmented system and T_{ref_i} being a SISO response of the PAV reference model. The number of frequencies was set to 20 and the frequency range was specified between $\omega_1 = 0.3$ rad/sec and $\omega_{n_\omega} = 16$ rad/sec.

3. The sum of the differences between the rotational velocities frequency responses of the augmented system and those of the PAV reference model, with each difference computed in terms of magnitude and phase as in Eq. 3.6.

The first objective function was defined to eliminate all solutions that could generate high control actions or cause instability. The second and the third might seem redundant objectives since the performance requirement are already included in the first objective function. However, they were necessary to prevent the optimization problem from choosing small performance gains to minimize γ.

This optimization problem defined in this way is a multi-objective one with competing objectives (the second and the third objectives compete with each other and with the first objective). This kind of problems can be solved through the application of genetic algorithms [Boyd and Barratt, 1991]. Therefore, a multi-objective genetic algorithm was implemented in Matlab to solve the optimization. The genetic algorithm provided a set of solutions called Pareto front [Boyd and Barratt, 1991]. Among these solutions those with $\gamma > 1$ were eliminated to ensure good disturbance rejection and satisfactory performance. Then, the optimal solution was selected as the one with minimum average value between second and third objective functions.

The resulting weighting functions obtained from the optimization problem are reported in Eqs. 3.7 and 3.8.

$$
W_u = diag\left[W_{u_{lat}}, W_{u_{lon}}, W_{u_{ped}}, W_{u_{col}}\right]
$$

$$
= diag\left[0.0283, 0.0635, 0.0010, 0.0225\right]\frac{0.05s+1}{0.005s+1}
$$

3.7

$$
W_p = diag\left[W_{p_p}, W_{p_q}, W_{p_r}, W_{p_u}, W_{p_v}, W_{p_w}\right] = diag\Big[9.0010, 9.7547,
$$

$$
6.1185, 5.7131, 1.0586, 1.0469\Big]\frac{0.005s+1}{0.05s+1}
$$

3.8

It is interesting to notice that the performance gains associated with the roll $[W_{p_p}]$ and the pitch axes $[W_{p_q}]$ have similar values and are approximately nine times larger than the longitudinal $[W_{p_u}]$ and the lateral $[W_{p_v}]$ gains, respectively.

The weighting functions in equations 3.7 and 3.8 were obtained for the H∞ control problem. However, they were also applied to the μ-synthesis control design as it was computationally harder to find ad-hoc optimal weights for this control technique.

3.4   Results

The H∞ and the μ-synthesis control design described in the previous sections provided controllers with orders 61 and 105, respectively. To reduce the controllers complexity, thus making them computationally less demanding, an order reduction was implemented based on the Hankel norm [Zhou and Doyle, 1999]. From the reduction, an H∞ controller with order 35 and a μ-synthesis controller with order 52 were obtained. The reduced controller were capable of achieving same stability and performance results as the original high order ones.

The assessment of the two reduced controllers will be now considered in terms of nominal and robust stability and nominal and robust performance. Furthermore, HQs evaluations will be done to compare the responses of the two augmented control systems with respect to the PAV reference dynamics.

3.4.1   Stability

Nominal stability was achieved by both implemented H∞ and μ-synthesis controllers. However, robust stability against model parametric uncertainties was achieved by the μ-synthesis control technique only, as can be seen in Figure 3.2. In this figure, lower and upper

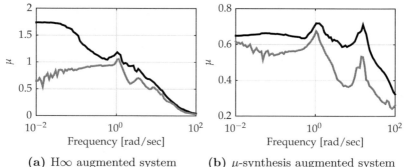

(a) H∞ augmented system   (b) μ-synthesis augmented system

Figure 3.2: Structured singular value (μ) lower and upper stability margins

bounds of the structured singular value μ are shown for the two augmented systems (helicopter model + controller). The H∞ augmented response reaches a maximum value of μ equal to 1.734. Therefore, the stability of the system can be preserved for uncertainties with infinity norm less than $1/1.734 = 0.58$, based on the theory presented in [Zhou and Doyle, 1999]. Since the infinity norm of the parametric uncertainty matrix is here equal to 1, the H∞ controller cannot ensure robust stability for all possible parametric uncertainties. For the μ-synthesis control technique, a maximum value of μ equal to 0.724 is obtained. Therefore, the stability of the system can be guaranteed against uncertainties with infinity norm less than $1/0.724 = 1.38$, which includes all considered parametric uncertainties.

3.4.2 Nominal Performance

The nominal performance of the implemented controllers was evaluated by considering the infinity norms of the nominal (without uncertainties) closed-loop augmented systems. For the H∞ augmented system, an infinity norm value equal to 0.865 was obtained, corresponding to good levels of disturbance rejection, noise rejection and satisfying reference response tracking. For the nominal μ-synthesis augmented system, an infinity norm value of 0.929 was obtained. Therefore, the H∞ controller achieved better nominal performance.

This result is confirmed in the frequency responses shown in Figures 3.3-3.4. The plots show the nominal (without uncertainties) frequency responses of the original identified helicopter model (Helicopter), of the two augmented control systems (H∞ and μ-synthesis) and of the reference model (PAV). The phases are represented in this way to allow for comparisons with those responses that present large delays. Figure 3.3 shows the controllers capability of modifying the original helicopter responses to track the reference model responses within the frequency range of interest [0.1-16 rad/sec]. However, a noticeable delay is introduced in the longitudinal and lateral axes by both controllers (Figures 3.3a-3.3b). A good fitting is achieved by both augmented systems in the vertical axis (Figure 3.4a) throughout the entire frequency range of interest. Also the yaw axis response (Figure 3.4b) shows a very good tracking, although the μ-synthesis augmented system performs well at low frequencies but diverges from the reference at higher frequencies.

Another comparison in terms of nominal performance can be made by considering the handling qualities analysis. Here, the ADS-33E-PRF [Baskett, 2000] short-term bandwidth (ω_{BW}) and phase delay (τ_p) parameters are considered to compare the augmented systems with respect to each other and to the reference PAV dynamics. In the

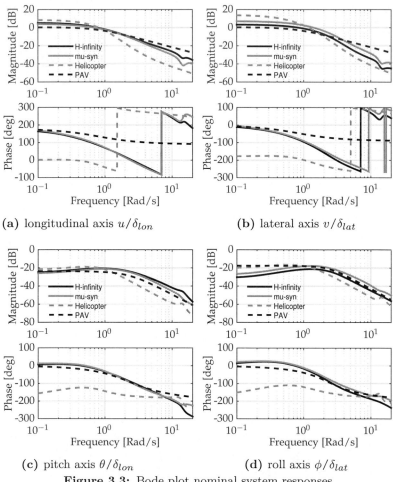

(a) longitudinal axis u/δ_{lon} (b) lateral axis v/δ_{lat}

(c) pitch axis θ/δ_{lon} (d) roll axis ϕ/δ_{lat}

Figure 3.3: Bode plot nominal system responses

ADS-33E-PRF three levels are defined to rate helicopters handling qualities: Level 1 corresponding to desirable handling characteristics, Level 2 corresponding to adequate handling characteristics, and Level 3 corresponding to undesirable handling characteristics.

As can be seen in Figure 3.5, the identified helicopter model shows adequate HQ performance (Level 2) in all axes (pitch, roll and yaw). The reference responses defined for PAVs have all Level 1 desirable HQ performance. From the plots it can be noticed how the augmented control systems try to resemble the HQs performance of the reference model in all three axes, by achieving Level 1 HQs. The two augmented control systems are comparable in the pitch axis (Figure 3.5a), whereas the H∞ performs better in roll and yaw axes (Figures 3.5b-3.5c). Therefore, also the HQs analysis confirms that a

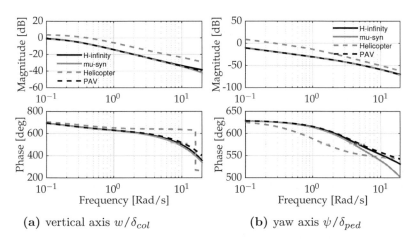

(a) vertical axis w/δ_{col} (b) yaw axis ψ/δ_{ped}

Figure 3.4: Bode plot nominal system responses

better nominal performance is achieved by the H∞ augmented system with respect to the μ-synthesis. However, some discrepancies can be seen in each axis between the two augmented systems and the PAV reference model. The magnitude of these discrepancies will be evaluated in the next chapter.

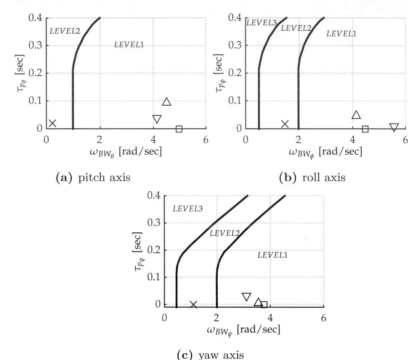

(c) yaw axis

Figure 3.5: ADS-33 nominal system; $\times$ = Helicopter; $\square$ = PAV; $\triangle$ = H-inf; ∇ = mu-syn

3.4.3   Robust Performance

The robust performance was evaluated through the μ-analysis, as considered for the robust stability. As can be seen in Figure 3.6, both augmented responses show a maximum upper bound of μ larger than 1. Therefore, both implemented controllers are not able to ensure robust performance against model parametric uncertainties.

The frequency responses of the perturbed augmented systems show how the dynamic responses are affected by different uncertainty values (Figures 3.7-3.8). The perturbed systems consists of 20 different responses obtained by randomly varying the values of the helicopter model parameters within the considered uncertainty boundaries (defined as $\pm 3CR_i$ for each identified parameter i). It can be noticed how the H∞ augmented system responses are particularly affected

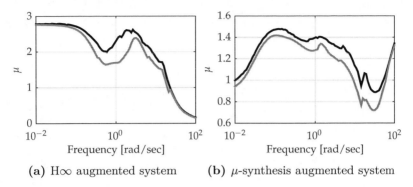

(a) H∞ augmented system   (b) μ-synthesis augmented system

Figure 3.6: Structured singular value (μ) lower and upper performance margins

by the uncertainties. Furthermore, for both augmented systems the vertical and yaw axes are less affected, as can be seen in Figure 3.8. These results will be discussed in the next section.

Figure 3.9 shows how the model uncertainties affect the HQs of the augmented systems. Also here 20 different responses are considered by randomly varying the uncertainty values. In the pitch and roll axes of both augmented systems, the uncertainties mainly affect the bandwidth (ω_{BW}) value. The yaw axis is almost not affected, as already noticed in the related frequency response (Figure 3.8b).

In Figure 3.9 the worst performance case is also shown, which corresponds to the maximum singular value of the uncertain frequency response matrix. It is interesting to notice that even the worst performance case falls within the HQs Level 1.

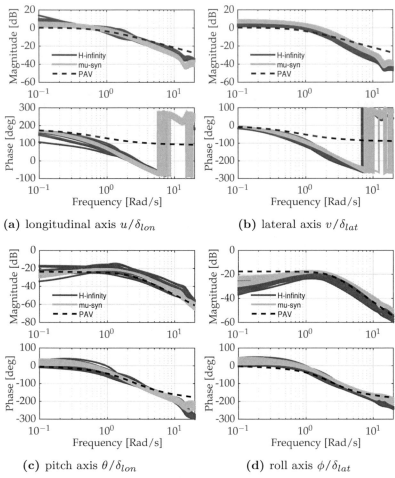

(a) longitudinal axis u/δ_{lon} (b) lateral axis v/δ_{lat}

(c) pitch axis θ/δ_{lon} (d) roll axis ϕ/δ_{lat}

Figure 3.7: Bode plot uncertain system responses

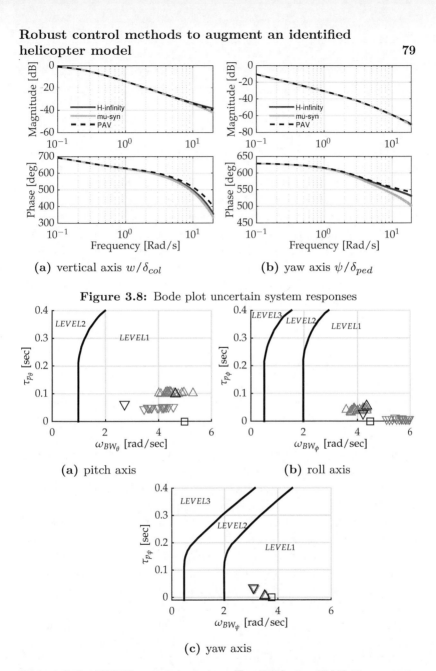

(a) vertical axis w/δ_{col}

(b) yaw axis ψ/δ_{ped}

Figure 3.8: Bode plot uncertain system responses

(a) pitch axis

(b) roll axis

(c) yaw axis

Figure 3.9: ADS-33 uncertain system; $\square$ = PAV; $\triangle$ = H-inf; $\triangledown$ = mu-syn; $\blacktriangle$ = H-inf worst case; $\blacktriangledown$ = mu-syn worst case

3.5   Discussion

The results presented in the previous sections allow for comparisons between the H∞ and the μ-synthesis augmented systems in terms of stability and performance requirements. The unachieved robust stability of the H∞ augmented system is an important point to be considered if an actual controller has to be implemented. This result is attributed to the fact that the H∞ method does not include any uncertainty information into the control design. Therefore, the controller resulting from this method can guarantee robust stability against disturbances and noise but not against model parametric uncertainties. On the other hand, the μ-synthesis control technique includes an uncertainty block structure into the design, giving the possibility to achieve robust stability.

Nominal performance was satisfactorily achieved. Both augmented control systems were able to resemble the PAV reference responses as far as allowed by the dynamic limits imposed by the helicopter model. The achievement of Level 1 HQs performances in all axes is an important result, as it gives the possibility to answer to the main question considered in this chapter: classical control techniques can augment an identified linear helicopter model to resemble HQs defined for PAVs.

Although Level 1 HQs were achieved in all axes by both augmented systems, an assessment of the observed discrepancy between the augmented systems and the PAV is not possible from these analyses as no definition exists of HQs levels for non-expert pilots. In the next chapter the experiment conducted for this purpose will be presented. In this experiment non-expert pilots performed piloted closes-loop control tasks in a simulated environment controlling the PAV reference model and the two augmented systems. Results will be evaluated in terms of performance and workload to assess the magnitude of discrepancy registered in the HQs analysis .

An important result is that the H∞ augmented system achieved better nominal performance than the μ-synthesis system. This result is attributed to the μ-synthesis system capability of achieving robust stability. In fact, worse performance capabilities can be related to a trade-off between robustness and performance. This also explains why the H∞ augmented system responses appeared more affected by model uncertainties (Figure 3.7). The weighting functions selection could be another reason for the worse nominal performance achieved by the μ-synthesis system. In the μ-synthesis control implementation the same weighting functions were used as those considered for the H∞ control method. A different optimization problem could be implemented to find weighing function parameters well suited for the μ-synthesis control method in order to verify if better performance can be achieved.

The robust analysis showed the controllers incapability of ensuring performance requirements against model uncertainties. Nevertheless, the worst performance case showed that both augmented systems are able to maintain Level 1 HQs, ensuring an improvement with respect to the Level 2 of the original helicopter model. Some control techniques (e.g. adaptive control methods) could be applied on the augmented systems to try to mitigate the effect of the model uncertainties on the nominal performance.

3.6   Conclusions

This chapter described the implementation of H∞ and μ-synthesis control techniques to augment an identified civil light helicopter model in hover. The resulting augmented systems were evaluated in terms of achieved robust stability, nominal performance and robust performance. A multi-objective optimization problem was implemented to select the parameters of the weighting functions used for the two control designs. The results showed that both considered classic control methods can modify the identified helicopter responses and handling qualities to resemble those defined for PAVs. Even though better nominal performance results were obtained with the H∞ control method, robust stability requirements were achieved by only the μ-synthesis control approach. However, both control techniques performed poorly for some specific uncertainty values, not ensuring performance robustness. Nevertheless, for both augmented systems the worst performance case fell within Level 1 HQs boundaries.

The results obtained in this chapter open the possibility for further studies that could be considered to expand and validate the control techniques implemented here. Control strategies could be investigated to mitigate the incapability of both augmented systems of ensuring required performance against model uncertainties.

The results presented in this chapter complete the second objective of the thesis. The next two chapters will implement the third objective and will go towards the direction of the studies proposed above.

CHAPTER 4

Evaluation and validation of augmented control systems in piloted closed-loop control tasks

This chapter presents the results of the experiment per-formed to evaluate the control augmented systems imple-mented in Chapter 3. Participants with no prior flight experience were asked to perform piloted closed-loop con-trol tasks in the MPI CyberMotion Simulator. Two maneuvers were selected: hover and lateral reposition. Four different dynamics were considered: the identified helicopter model, the two augmented systems and the reference PAV model. The four dynamics were compared in terms of objective and subjective workload and perfor-mance. Results presented in this chapter provide a better understanding of the implemented augmented systems and shed light on the magnitude of discrepancy with respect to each other, to the original identified helicopter model and to the PAV reference model.

The contents of this chapter are based on:

Paper title	Validation of Augmented Civil Helicopters in the MPI CyberMotion Simulator
Authors	S. Geluardi, J. Venrooij, M. Olivari, L. Pollini and H. H. Bülthoff
In preparation	Journal of Guidance, Control, and Dynamics

I N the past few years the myCopter project has investigated new con cepts and technologies for alleviating congestion issues that affect the road network of major cities all over the world [Nieuwenhuizen et al., 2011]. A radical solution has been proposed that consists of realizing a Personal Aerial Transport System (PATS) to combine the best of ground- and air-based transportation features.

The key element of PATS would be the Personal Aerial Vehicle (PAV) that the traveling public would use. Although designing new vehicle prototypes was not among the project's goal, it was considered important to define response types and Handling Qualities (HQs) that should enable an average car driver to fly a PAV [Perfect et al., 2015a]. To determine what kind of response types were suitable for "inexperienced pilots" (pilots with no prior flight experience), flight tests were performed in the HELIFLIGHT-R flight simulator at the University of Liverpool. Results showed that Translational Rate Command (TRC) response types for longitudinal, lateral and vertical control axes and Rate-Command Attitude-Hold (RCAH) response types for the yaw control axis allow inexperienced pilots to execute demanding tasks with good performance. This response types were tuned to ensure Level 1 Handling Qualities according to the requirements of the Aeronautical Design Standard-33 (ADS-33E-PRF) [Perfect et al., 2015a].

Starting from these results, in this thesis it has been proposed to consider civil light helicopters as candidates for PAVs. Chapter 3 showed that augmented control systems can be applied to helicopter models to resemble dynamics and HQs of a PAV even against atmospheric turbulences and noise. However, some discrepancies were found between the nominal control augmented systems (H∞ and μ-synthesis) with respect to each other and with respect to the PAV reference model. To assess the magnitude of these discrepancies and to shed light on capabilities and limitations of the implemented control systems, an experiment was designed. In this experiment, participants without prior flight experience were invited to perform piloted closed-loop control tasks in a simulated environment. The experiment was performed at the Max Planck Institute for Biological Cybernetics and the MPI CyberMotion Simulator was used. Two maneuvers were selected: hover and lateral reposition. Four experimental dynamics were considered: the identified helicopter model, the two augmented systems and the reference PAV model. The four dynamics were evaluated and compared in terms of subjective and objective workload and performance.

This chapter presents the results of the experiment. The chapter is structured as follows. First, the experimental method and the implemented setup are described in details. Then, the obtained results are presented with a final discussion. Conclusions and recommendation are given at the end of the chapter.

4.1   Experimental method

4.1.1   Apparatus

The experiment was performed in the CyberMotion Simulator (CMS), shown in Figure 4.1a. The CMS is a motion simulator based on an anthropomorphic robot, manufactured by KUKA Roboter GmbH. This 8 degrees of freedom robotic arm allows for highly realistic flight scenarios to be simulated as a result of its high dexterity and the large motion envelope. The end-effector of the arm is an enclosed cabin with a large field-of-view that allows for virtual environments to be projected (Figure 4.1b). In this experiment the cabin was equipped with a pilot seat, a conventional center-stick cyclic, a collective lever and rudder pedals (Wittenstein GmbH, Germany). Stiffness and damping force of each control axis were tuned with the help of a helicopter pilot till a realistic and natural feel was obtained.

A motion cueing algorithm (MCA) was used to control the simulator within its motion envelope. The selected MCA was based on washout filters able to convert the motion of the simulated vehicle into simulator input commands. The adopted washout filters were second-order high-pass filters and their gains were manually tuned based on evaluations given by one helicopter pilot (for the helicopter dynamics) and three non-experienced pilots (for the augmented dynamics). This, in order to obtain a motion coherent with the visual cues provided in the experiment.

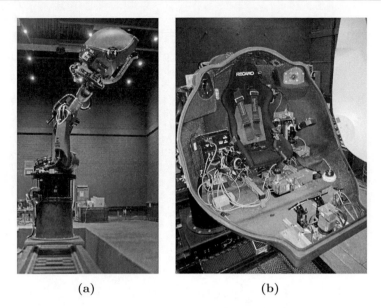

(a)   (b)

Figure 4.1: Sub-figure (a) shows the MPI CyberMotion Simulator. Sub-figure (b) shows the internal of the simulator cabin equipped with a pilot seat, a center-stick cyclic, a collective lever and rudder pedals.

The visual scenery used for the experiment (Figure 4.2) was implemented with the game development system Unity [Uni, 2015]. The scenery presented on the inside of the simulator cabin was based on the hover and the lateral reposition maneuvers described in the ADS-33E [Baskett, 2000]. The projectors used inside the cabin had an update rate of 60 Hz. The time delay of the visual cues was measured in a previous experiment to be approximately 40 ms [Wiskemann et al., 2014].

4.1.2 Independent variables

The experiment was designed to evaluate four different dynamics in piloted closed-loop control tasks performed by inexperienced pilots. The four dynamics represent the independent variables of the experiment. The first dynamics is the identified model of a civil light

Figure 4.2: Unity3D visual of airport with ADS-33E-PRF features for MTEs.

helicopter (R44 Robinson Raven II) in hover condition (Chapter 2). This dynamics was considered as a baseline to assess inexperienced pilots capabilities to perform demanding maneuvers, while controlling a model characterized by instability, non-minimum-phase and high control coupling.

The second and the third dynamics are the two augmented control systems (H∞ and μ-synthesis) described in Chapter 3. These dynamics were considered to evaluate discrepancies with respect to each other and with respect to the PAV reference model.

The fourth dynamics is the PAV model implemented during the myCopter project by University of Liverpool [Perfect et al., 2015b]. This dynamics is a simplified decoupled helicopter model with a transfer function representation of the HQs suited for inexperienced pilots. This dynamics was selected as a reference to allow for comparisons with the augmented control systems.

4.1.3   Mission Task Elements

To evaluate the considered dynamics, two different maneuvers were selected: the hover and the lateral reposition maneuvers. Both maneuvers, also called Mission Task Elements (MTE), are defined in the

ADS-33E-PRF [Baskett, 2000] but were adapted here according to the purpose of the experiment. For example, the validity of the identified helicopter model within a velocity of 20 knots (10.29 m/s) did not allow the MTEs to be performed with the time limits defined in the ADS-33E-PRF. Therefore, time limit constraints were not taken into account in the experiment. Furthermore, definitions of "desired" and "adequate" given in the ADS-33E-PRF were not adopted here. These definitions are used to rate experienced helicopter pilots capabilities at performing MTEs. However, in this experiment participants were not requested to achieve same performance as highly trained helicopter pilots.

The adapted MTEs were defined as follows.

- Hover MTE. This maneuver starts in a stabilized hover at an altitude of 25 ft (7.62 m) in front of a green sphere. The target is oriented 45 degrees right, relative to the heading of the vehicle. The pilot starts the maneuver by following a white line on the ground till the final target is reached. A green square on the ground locates the target position with an accuracy of ±10 ft (±3.048 m). Furthermore, an outer yellow border is used to locate the target position with an accuracy of ±15 ft (±4.57 m). After reaching the target, a stabilized hover is to be maintained for 30 seconds.

  To aid the pilot in maintaining the right position during the hover phase, a hover board is provided 150 ft (45.72 m) in front of the target position. Furthermore, a pole high 50 ft (15.24 m) with a red sphere on top is placed half way between the location of the target and the board to aid the pilot in maintaining the correct vertical and lateral position. Finally, a horizontal white line is placed on the ground to aid the pilot in maintaining the correct longitudinal position.

- Lateral Reposition MTE. This maneuver starts in a stabilized hover at an altitude of 25 ft in front of a green sphere. The initial longitudinal axis of the vehicle is oriented 90 degrees left with respect to the target position. A lateral acceleration followed by a deceleration is to be performed to reach the target,

300 ft (91.44 m) right with respect to the initial position. Then, a stabilized hover is to be achieved and maintained for 5 seconds. During the entire maneuver, initial vertical position and heading have to be maintained.

A green square on the ground locates the target position with an accuracy of ±10 ft, while an outer yellow border is used to locate the target position with an accuracy of ±15 ft. A straight white line on the ground is used to guide the pilot from the initial position to the target. Green and yellow markings are placed on the ground throughout the path to respectively indicate ±10 ft and ±15 ft distances from the straight white line. Throughout the path black poles 25 ft high and with a spacing of 50 ft are placed in front of the vehicle. These poles aid the pilot in maintaining the correct vertical position during the entire maneuver.

In the hover maneuver the longitudinal axis is particularly excited during the first phase in which the target position has to be achieved. Therefore, the capability to control this axis in low speed condition can be evaluated. Conversely, the hover phase allows for an evaluation of the capability to maintain a stabilized zero velocity condition, which is directly influenced by the control axes-coupling and the vehicle handling qualities.

In the lateral reposition maneuver the lateral control axis is mainly excited allowing for its evaluation in low speed condition. Furthermore, requirements concerning maintaining same attitude and height throughout the entire maneuver can be used to evaluate the capability to control the yaw and the vertical axes in low speed flight regime. Therefore, the two maneuvers were selected to allow for all control axes to be evaluated in both zero and low speed flight regimes.

4.1.4   Dependent measures

To evaluate differences among the four dynamics selected as independent variables, four main metrics were considered as dependent measures: objective performance, objective workload, subjective performance and subjective workload.

Objective perfomance metrics were defined as "position error" and "time to accomplish an MTE" that subjects had to minimize during each trial. For the hover MTE, the position error was defined as the absolute difference between the target position and the vehicle position averaged over the 30 seconds of stabilized hover. The position error was computed in terms of longitudinal, lateral, vertical and heading errors. The time to accomplish the hover MTE was computed from the beginning of the trial till the hover target position was achieved.

For the lateral reposition MTE, the position error was defined as the absolute difference between the vehicle position and the reference position averaged over the time to accomplish the maneuver. The reference position was represented with a straight white line on the ground for the longitudinal axis and with ten vertical poles positioned throughout the path for the vertical axis. The reference for the yaw axis was defined by the initial heading. The lateral error was computed as the absolute difference between the target position and the vehicle position averaged over the last 5 second, during which a stabilized hover above the target had to be maintained. The time to accomplish the lateral reposition MTE was computed from the beginning of the trial till the final target position was achieved.

The objective workload was considered for both MTEs in terms of "amount of control activity", as defined during the myCopter project by University of Liverpool [Perfect et al., 2015a]. The amount of control activity represents the number of discrete control movements over the time necessary to complete the task. Discrete control movements are individual control deflections measured between two points where the control velocity is zero. In this experiment, control deflections smaller than one degree were discarded to prevent measurement noise. The number of control deflections per second (control input rate) was first computed for each control axis (longitudinal stick, lateral stick, collective and pedals). Then, the four control input rates were averaged to produce a single value per trial.

Subjective metrics of workload and performance were also considered to have a comparison with the objective measures. The NASA Task Load Index (TLX) was adopted for this purpose since it allows for six workload aspects to be evaluated: mental demand, physical demand, temporal demand, performance, effort, and frustration [Hart

and Staveland, 1988]. During the experiment, subjects were asked to
rate each workload aspect individually. Then, they had to compare
the aspects with each other to evaluate their contribution to the
overall workload. The obtained weighted ratings were later used to
compute for each MTE a single TLX workload score between 0 and
100, with lower numbers indicating lower workload.

4.1.5 Participants and instructions

Twenty-nine subjects were invited to participate in the experiment.
Nineteen subjects were male and ten female. Their ages ranged from
24 to 39 years old. All participants were inexperienced pilots except
one. The helicopter pilot had an experience of 110 flight hours and
more than 500 take-offs and landings. His last flight training was
in 1992. Most of his training was done with an Alouette II Sud-Est
SE.3130 helicopter. He also trained for 4 hours in a simulator with a
Bell UH-1D model.

Before entering the CMS, subjects received an extensive briefing
on the objective of the experiment (Appendix D). Furthermore, a
brief explanation was given about general helicopter dynamics and
helicopter control devices. Finally, subjects were instructed on the
MTEs to be performed.

The main instruction participants received was to minimize the
time to accomplish the maneuver and the position error in both MTEs
(hover and lateral reposition).

4.1.6 Experimental procedure

The experiment was designed as a "between subjects" one. This
means that each participant was invited to fly only one of the four
experimental dynamics. The helicopter pilot was assigned to the
experimental condition involving the control of the identified helicopter model. This, in order to allow for comparisons with respect
to the other participants. The other subjects were assigned to an
experimental dynamics randomly selected among the four considered
in the experiment: identified helicopter model, H∞ augmented system, μ-synthesis augmented system and PAV model. In this way, the

inexperienced pilots were divided into 4 groups of 7 subjects each. Every participant was invited to fly both MTEs (hover and lateral reposition). Furthermore, the order of the MTEs was randomized to cancel potential learning effects, allowing for possible differences to be investigated.

The experiment was divided into two main phases. The first phase started with 20 minutes of training on the CMS. For each participant, at the beginning of the training one of the two MTEs was randomly selected. Then, a stable and decoupled dynamics was assigned to familiarize with the control devices and to start practicing on the selected MTE. After 5 minutes of training, one among the four dynamics was assigned. Subjects kept on practicing with the selected dynamics till the end of the training. During the training, subjects were constantly instructed by the experimenter via headphones. At the end of each trial, a score was provided for motivating participants to improve their performance and to maintain a constant level after their proficiencies had stabilized. The performance was displayed as longitudinal, lateral, vertical and heading errors and as time to complete the trial. At the end of the training, subjects had to rate the NASA TLX. After the training, subjects were invited to perform five final trials involving the same MTE selected in the training. The data of each trial were logged at a frequency of 100 Hz for later analysis. Finally, subjects were asked to rate again the NASA TLX for consistency with the evaluation given after the training. This concluded the first phase.

In the second phase, subjects were instructed on the second MTE to be performed. Then, a brief training (~ 10 minutes) was performed on the CMS. This training had shorter duration because participants had to practice on a different MTE but with the same control devices and the same dynamics assigned in the first experimental phase. After the training, participants had to rate the NASA TLX. Then, they were invited to perform five final trials involving the same MTE selected in the second training. The data of each trial were logged at a frequency of 100 Hz for later analysis. Finally, subjects were asked to rate again the NASA TLX for consistency with the evaluation given after the second training. This concluded the experiment. Each subject completed the experiment in approximately 2 hours.

The duration of the first training phases was established based on a pilot experiment, in which three inexperienced pilots were asked to perform the two selected MTEs, while controlling the PAV reference model. Results showed that after 20 minutes all participants were able to achieve adequate performance. Therefore, the same amount of training was assigned to all conditions for comparisons with the PAV one.

4.2 Hypotheses

Three main hypotheses were tested in this experiment. The first hypothesis involves the helicopter dynamics obtained from the identification of a civil light helicopter in hover condition (Chapter 2). The duration of the training in the experiment was hypothesized not sufficient compared to the several hours of training that an average person would need in order to learn how to stabilize and control a helicopter in hover. Therefore, inexperienced pilots who were assigned to this experimental condition were expected not to be able to perform the selected MTEs.

The second hypothesis concerns the PAV dynamics defined by University of Liverpool during the myCopter project. This dynamics was implemented to allow inexperienced pilots to perform demanding maneuvers in hover and low speed conditions. Therefore, participants who received this experimental condition were expected to be able to control the model with adequate performance, after the first training phase.

The third hypothesis concerns the two augmented dynamics implemented in Chapter 3, H∞ and μ-synthesis augmented systems. These dynamics were able to resemble response types and handling qualities of the PAV reference model. Nevertheless, some discrepancies were observed between the nominal augmented systems and the PAV model. Furthermore, the H∞ augmented system achieved slightly better performance results than the μ-synthesis one, as shown in the handling qualities plots of Figure 3.5. From this results, it was hypothesized that a decrease in performance would be observed in the μ-synthesis experimental condition. In particular, the hover

MTE was expected to be affected as the considered handling qualities results were obtained for the hover flight regime. No other hypotheses were developed concerning the magnitude of discrepancy between the augmented dynamics and the PAV dynamics as similar experiments had never been conducted before.

4.3 Results

In this section results are presented of the 29 subjects who participated in the experiment. Two participants assigned to the helicopter dynamics condition abandoned the experiment during the first training phase because of motion sickness. This was due to the incapability of stabilizing the helicopter dynamics, which induced abrupt and uncontrolled movements. Therefore, these participants were excluded from later analyses.

In the helicopter dynamics group only one inexperienced subject was able to perform the two MTEs. The other participants of the group were not able to control the helicopter model until the end of the experiment. This confirms the first hypothesis according to which 20 minutes of training are not sufficient to allow an ordinary inexperienced pilot to learn how to fly a helicopter. Conversely, all subjects who received the PAV dynamics were able to perform both MTEs, which confirms the second hypothesis.

In the next subsections the four selected dynamics will be evaluated and compared to each other. Statistical analyses will be performed on the collected data. Results will be presented through the use of boxplots.

4.3.1 Hover MTE objective performance and workload

In the helicopter dynamics group all subjects but one were not able to perform the hover maneuver and to reach the target position. Therefore, they were excluded from the following analyses since an adequate comparison with the other participants could not be done. It is reminded that the objective performance was defined as "position error" (longitudinal, lateral, vertical and heading) and "time to accomplish

the MTE". The analysis of the objective performance was made as
follows. First, the data collected in the last 5 trials were averaged to
obtain a single value for each participant. From these values, median
and standard deviation were computed for each of the four groups
defined in the experiment. The results can be visually seen in the
boxplots of Figures 4.3 and 4.4. In each box the central mark represents the median, while the edges are the 25th and 75th percentiles,
respectively. The maximum and minimum values correspond to ± 2.7
times the standard deviation, respectively. Outliers are individually
shown with a cross. A multivariate analysis of variance (MANOVA)
followed by a Bonferroni post-hoc test were considered to check for
statistical differences in longitudinal, lateral, vertical and heading
errors, between the different experimental groups. From the test, no
statistical difference was found between the H∞ augmented dynamics
and the PAV reference model. The μ-synthesis augmented system
performed slightly worse in longitudinal, lateral and vertical axes, as
can be seen in Figure 4.3. This confirms the third hypothesis. How-

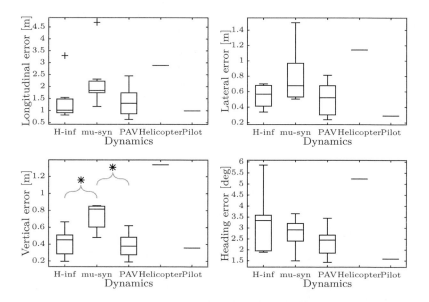

Figure 4.3: Position error hover MTE

ever, a statistically significant difference was observed in the vertical error only. In particular the μ-synthesis performed statistically worse than the H∞ with a p-value of 0.004 and statistically worse than the PAV reference dynamics with a p-value of 0.002. Nevertheless, it can be noticed that the vertical error difference between the medians is less than half a meter.

Figure 4.3 also shows the performance of the actual helicopter pilot and the one of the only inexperienced subject who was able to perform the maneuver in the helicopter dynamics group. As can be seen, the inexperienced subject of the helicopter group performed worse than the median of the other groups of participants. Conversely, the actual helicopter pilot achieved better performance than the medians of all other groups.

An analysis of variance (ANOVA) test was done on the time to accomplish the maneuver. No significant difference was found between the H∞, the μ-synthesis and the PAV dynamics (Figure 4.4a). Furthermore, it can be noticed how the median of the PAV group is very close to the value of the actual helicopter pilot.

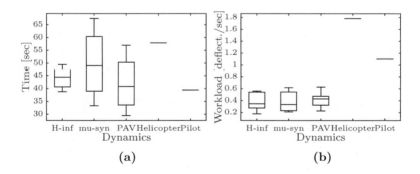

(a)   (b)

Figure 4.4: Sub-figure (a) shows the time to accomplish the hover MTE. Sub-figure (b) shows the objective workload measured as the averaged number of controls deflections per second

An ANOVA test was also performed on the objective workload defined as the amount of control activity. No statistical difference was found between the two augmented systems and the PAV reference model (Figure 4.4b). In the same figure it is possible to notice that a

larger amount of control activity was necessary to perform the hover
MTE for both the inexperienced subject of the helicopter group and
the actual helicopter pilot.

4.3.2 Hover MTE subjective rating

The boxplot obtained from the data collected with the NASA TLX
questionnaire is shown in Figure 4.5. Here, the data of all 29 partici-

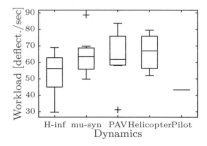

Figure 4.5: NASA TLX subjective workload hover MTE

pants are considered. The ANOVA test did not show any statistical
difference between the augmented systems and the PAV reference
model. This confirms the result obtained in the objective workload
measure. However, it can be noticed that subjects of the helicopter
dynamics group provided ratings comparable to those given by the
other participants. Moreover, the actual helicopter pilot rated his
overall workload lower than the medians of all other groups. These
results are in contrast with those obtained in the objective workload
measure and will be later discussed in the discussion section.

4.3.3 Lateral reposition MTE objective performance
and workload

In the lateral reposition MTE, the position error was computed since
the beginning of the maneuver. All participants were able to fly for
at least a few seconds. Therefore, a comparison of the collected data
was possible by including all subjects. Figure 4.6 shows the objective
performance as longitudinal, vertical and heading error. The lateral

axis is not shown because none of the participants in the helicopter group could reach the target position except one. Therefore, no comparisons was possible in this axis. As can be seen in the figure, subjects of the helicopter group performed significantly worse than all other groups of participants. These results further confirm the first hypothesis.

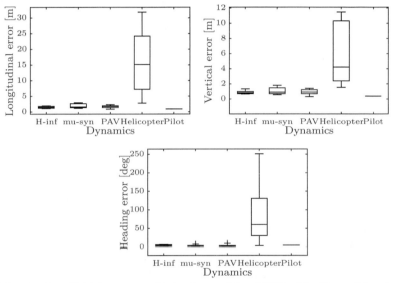

Figure 4.6: Position error of lateral reposition MTE with all participants

Participants of the helicopter group unable to reach the target position were excluded from any statistical analysis to avoid biased comparisons. For the objective performance measure, the statistical analysis was performed as follows. First, the data collected in the last 5 trials were averaged to obtain a single value for each participant. From these values, median and standard deviation were computed for each experimental group. The boxplots obtained from these data are shown in Figure 4.7 for longitudinal, lateral, vertical and heading errors.

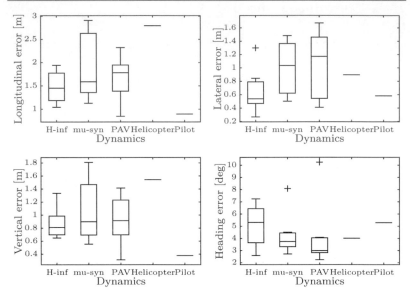

Figure 4.7: Position error lateral reposition MTE

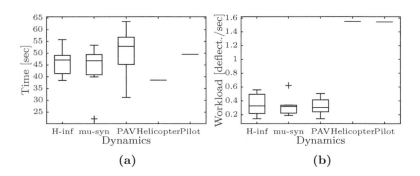

Figure 4.8: Sub-figure (a) shows the time to accomplish the lateral reposition MTE. Sub-figure (b) shows the objective workload measured as the averaged number of controls deflections per second

A multivariate analysis of variance (MANOVA) followed by a Bonferroni post-hoc test were performed to check for statistical differences in longitudinal, lateral, vertical and heading errors, between the different experimental groups. From the test, no statistical difference was found between the two augmented dynamics groups and the PAV group. As shown in Figure 4.7, the performance of the inexperienced subject in the helicopter dynamics group improved in the lateral and in the heading axes with respect to the hover MTE. However, longitudinal and vertical errors remained larger than the medians of all other groups. The actual helicopter pilot performed better than the medians of all other groups of participants only in the longitudinal and in the vertical axes.

An ANOVA test was performed on the time to accomplish the lateral MTE. No statistical difference was found between the augmented system dynamics groups and the PAV dynamics group (Figure 4.8a). Furthermore, the inexperienced subject of the helicopter dynamics group performed better than the medians of all other groups of participants.

An ANOVA test was also performed on the objective workload defined as the amount of control activity. No statistical difference was found between the two augmented systems and the PAV reference model (Figure 4.8b). Also in this MTE, a larger amount of control activity was necessary to perform the maneuver for both the inexperienced subject of the helicopter group and the actual helicopter pilot.

4.3.4   Lateral reposition MTE subjective rating

The boxplot obtained from the data collected with the NASA TLX questionnaire is shown in Figure 4.9. Here, the data of all 29 partic-

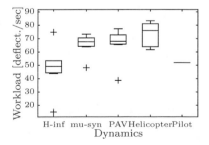

Figure 4.9: NASA TLX subjective workload lateral reposition MTE

ipants are considered. As can be seen in the figure, subjects of the Hoo group rated their overall workload lower than all other groups of participants. Furthermore, subjects of the helicopter dynamics group gave similar subjective workload ratings as the participants of the μ-synthesis and the PAV groups. The rating of the actual helicopter pilot was very close to the median of the Hoo group and lower than the medians of all other groups of participants. These results are again in contrast with those obtained from the objective workload measure and will be discussed in the next section.

4.4 Discussion

The results of the experiment provided insights into the four investigated dynamics. In particular, the chosen MTEs allowed for evaluations and comparisons of the augmented systems (Hoo and μ-synthesis) with respect to the identified helicopter dynamics and to the PAV reference model. The first important result was that 6 out of 7 inexperienced participants of the helicopter group were not able to achieve stability and to perform the selected MTEs. This result confirmed the first hypothesis according to which 20 minutes of training are not sufficient to allow an ordinary inexperienced pilot to learn how to fly a helicopter. However, one inexperienced subject of the helicopter group was able to perform both maneuvers after the first training phase. This result was attributed to the previous subject experience with control tasks involving helicopter-like dynamics in simulation. Furthermore, the subject was able to quickly learn how to control the dynamics by taking advantage of the adopted training approach. This approach was considered more effective than the one generally used by helicopters instructors since subjects had the possibility to loose control and crash. As such, they could learn from their mistakes and improve their performance after each trial. Clearly, this would not be possible in a real scenario. Although this approach was beneficial for only one participant in the helicopter dynamics group, it opens the possibility for further research. For example, it could be investigated which factors are important during the learning process or whether few hours of flight tests in a motion simulator can be beneficial before training on an actual helicopter.

The second important result was that all participants of the PAV group and those of the two augmented systems were able to correctly perform the selected maneuvers. This result confirmed the second hypothesis, provided insights into the features of the augmented systems and gave information about the magnitude of discrepancy between the augmented systems and the PAV reference model. Objective performance and workload measures showed no significant difference between the H∞ augmented system and the PAV reference model in both MTEs. The μ-synthesis augmented system performed equally well as the PAV reference model in the lateral reposition MTE. However, a decrease in performance was observed in the hover MTE. Although this result confirmed the third hypothesis, a significant difference was observed only in the vertical axis with a difference of half meter between the median of the PAV reference model and that of the μ-synthesis augmented system (Figure 4.3).

From these results, it can be concluded that both augmented systems can resemble PAVs response types and handling qualities in piloted closed-loop control tasks. This positively answers the main question of the thesis as to whether a helicopter can be transformed into a PAV. Therefore, the approach proposed here represents a valid alternative to the possibility of implementing new vehicles with handling qualities well suited for PAVs pilots.

The third important result of the experiment was that the actual helicopter pilot achieved higher performance than the medians of all other groups of participants. This was expected, given the pilot's previous experience with actual helicopters and flight simulators. However, it was not possible to exclude beforehand whether subjects of the PAV dynamics group would achieve similar or better performance than the actual pilot. As a matter of fact, some participants of the H∞ and the PAV groups performed as good as the actual pilot in the hover MTE. Nevertheless, this result showed that ordinary PAVs pilots would need longer trainings than the one considered in this experiment to achieve performances comparable to those of experienced helicopter pilots.

The final interesting result was provided by the NASA TLX questionnaire as participants ratings on subjective workload and performance provided different results than those obtained from the objective measures. As presented in Figure 4.9, subjects of the H∞ dynamics group rated their workload significantly lower than all other participants. Furthermore, no significant difference was found between all other groups in the lateral MTE. Comparable ratings were given by all groups of participants also in the hover MTE. Finally, in both MTEs the actual helicopter pilot provided subjective ratings in contradiction with the results obtained from the objective measures.

These results were attributed to the design of the experiment, which was conceived as a "between subjects" one. This means that every participant had the chance to practice and perform maneuvers with only one of the four possible dynamics. As a result, they rated the workload associated to the assigned experimental dynamics, without having any reference for comparisons. Therefore, subjects focused on different aspects like the difficulty at controlling simultaneously four control devices while flying in a 3 dimensional space, something they had never done before except for the actual pilot. Many inexperienced pilots considered the two MTEs as highly demanding tasks. Conversely, the actual pilot did not find any particular difficulty at performing the selected maneuvers. Because of these discrepancies, the NASA TLX questionnaire did not provide additional findings to evaluate the hypotheses and to answer the questions considered in this thesis.

4.5 Conclusions

This chapter presented the results of the experiment performed in the MPI CyberMotion Simulator. The aim was to assess the implemented control augmented systems described in Chapter 3. In the experiment piloted closed-loop control tasks were performed by pilots with no prior flight experience. Two control tasks maneuvers were considered: the hover and the lateral reposition MTEs. Each participant was asked to perform both MTEs with one dynamics among the identified helicopter model, the PAV reference model, the H∞ augmented system and the μ-synthesis augmented system. An actual helicopter pilot was also invited to perform the two MTEs with the helicopter dynamics. This, to allow for comparisons with the other participants.

At the beginning of the chapter the experimental method and the implemented setup were described in details. Then, three main hypotheses were considered to be verified in the experiment. The results were evaluated in terms of objective and subjective workload and performance and were presented through the use of boxplots. At the end of the chapter, a final discussion on the main experimental findings was given.

The main conclusion of this experiment was that both implemented augmented systems (H∞ and μ-synthesis) were able to achieve PAVs performance and workload in piloted closed-loop control tasks. This result positively answered the main question asked in this thesis as to whether civil helicopters can be considered as possible candidates for PAVs.

This chapter completes the third and last objective of the thesis. In the next chapter, a solution is proposed to reduce the detrimental effect of the model parametric uncertainties on the performance of the augmented systems implemented in Chapter 3.

CHAPTER **5**

$\mathcal{L}_1$-based model following control to suppress model parametric uncertainties effects

Aim of this chapter is to propose a solution to mitigate the detrimental effect of helicopter model parametric uncertainties on the performance of H∞ and μ-synthesis augmented systems presented in Chapter 3. An $\mathcal{L}_1$-based adaptive control is considered for this purpose. To verify the efficacy of this approach, the adaptation is applied to a simple PID-based control. Furthermore, only the translational rate commands defined for PAVs are considered as reference dynamics. First, a PID-based control is implemented to augment the nominal helicopter model without uncertainties. Then, an $\mathcal{L}_1$ adaptive controller is designed to restore the nominal responses of the augmented helicopter when parametric uncertainties are added. Results show that the application of the adaptive controller on the augmented helicopter dynamics can significantly reduce the effects of model parametric uncertainties. Therefore, further studies could be considered to apply the $\mathcal{L}_1$-based adaptive control to the H∞ and the μ-synthesis augmented systems.

The contents of this chapter are based on:

Paper title	$\mathcal{L}_1$-based Model Following Control of an Identified Helicopter Model in Hover
Authors	G. Picardi, S. Geluardi, M. Olivari, L. Pollini, M. Innocenti and H. H. Bülthoff
Published in	Proceedings of the American Helicopter Society 72th Annual Forum, Palm Beach, Florida, 2016

THE Personal Aerial Vehicle model defined during the myCopter
project has been considered throughout this thesis as a reference for civil light helicopters to make them accessible to the general
public. In Chapter 3 it was proved that optimal (H_∞) and robust
(μ−synthesis) control methods can augment helicopters to achieve
this goal. In particular, good results were obtained in the nominal
case of exact knowledge of the helicopter parameters (Figure 3.5).
Furthermore, these augmented systems were validated through piloted closed-loop control tasks, performed in the MPI CyberMotion
Simulator by inexperienced pilots (Chapter 4). Results showed that
both augmented systems are able to achieve comparable results to
the PAV reference model in terms of performance and workload. The
experiment did not considered the effects of parametric uncertainties
to allow for comparisons between the considered dynamics. However,
in Chapter 3 it was shown that both augmented systems cannot ensure
nominal performances when parametric uncertainties are added to
the helicopter model (Figure 3.9).

This performance degradation was attributed to the conservative-
ness of robust control laws that are designed to handle the worst
parametric uncertainty case. As a result of this, control system performance might be sacrificed to achieve a desired degree of robustness. To
overcome this limitation, adaptive control methods can be considered.
Adaptive controllers perform an online estimation of the uncertainties
and produce control inputs able to reduce the undesirable deviations
of the uncertain system from the nominal behavior [Lavretsky and
Wise, 2012].

Many successful applications of adaptive control can be found in
the field of small scale helicopter control. In the work of Guerreiro
$\mathcal{L}_1$-adaptive control theory was used to provide attitude and velocity
stabilization of an autonomous small scale rotorcraft [Guerreiro et al.,
2009]. The adaptive controller was designed to reject the effects of
wind disturbances. In this thesis, however, the focus is on a full scale
manned helicopter and the adaptive controller has to be designed to
reject the effects of parametric uncertainties due to identification. In
the work of Bichlmeier et al. [Bichlmeier et al., 2013] an $\mathcal{L}_1$-adaptive
controller was designed to augment a preexisting Proportional Integral
Derivative (PID) baseline controller on a full scale manned helicopter.

The aim of the adaptive controller was to maintain handling qualities in situations the baseline controller was not designed for or performed poorly. The adaptive controller compensated for uncertainties that enter the system dynamics through the control channel, i.e. *matched* uncertainties. Nevertheless, the identified model considered in this thesis is affected by uncertainties that do not fall into the matched category, the so called *unmatched* uncertainties. Few works in literature refer to adaptive control with unmatched uncertainties. One example is given in [Gregory et al., 2009], where the $\mathcal{L}_1$-adaptive control was applied to the NASA AirStar aircraft. However, to the best of the writer's knowledge, adaptive control has never been applied to full scale helicopter models to reject the effects of unmatched uncertainties.

For these reasons, the goal of this chapter is to investigate whether adaptive control methods can be successfully used to compensate for detrimental effects of parametric uncertainties on an augmented helicopter model. To verify the efficacy of this approach, the adaptation is applied to a simplified problem, in which a PID-based controller is used to achieve stability and to follow the PAVs reference responses. Furthermore only PAVs translational rate command (TRC) responses are selected as reference dynamics. Aim of the adaptive controller is to restore the nominal behavior of the augmented helicopter model when uncertainties are added.

The chapter is organized as follows. First, a brief description of the identified helicopter model is given. Second, the implementation of the PID-based controller is presented. Third, the theory and the implementation of the considered adaptive control is described. Finally, a Monte-Carlo study is conducted and results are shown to assess the proposed design.

5.1 Helicopter Model and Augmentation System

5.1.1 Helicopter Model

Here, the identified state-space model presented in Chapter 2 is considered. This model depends on 28 parameters and a minimum realization is considered here with 4 inputs, 8 outputs and 17 states.

The model inputs are the cyclic stick $[\delta_{lon}, \delta_{lat}]$, the collective (δ_{col}) and the pedals (δ_{ped}) deflections. The outputs considered in this Chapter are the translational velocities $[u, v, w]$, the angular rates $[p, q, r]$, roll and pitch angles $[\phi, \theta]$. It is reminded that the model contains important rotor-fuselage coupling dynamics such as flapping, inflow and lead-lag.

The resulting open loop helicopter dynamics can be written as:

$$\dot{x} = A_{he}(\rho)x + B_{he}(\rho)u$$
$$y = C_{he}x$$

5.1

where $x \in \mathbb{R}^{17}$ is the state vector, $u \in \mathbb{R}^4$ the input vector and $y \in \mathbb{R}^8$ the output vector of the system. The matrices A, B, C describe the system dynamics and the vector $\rho \in \mathbb{R}^{28}$ contains the identified parameters.

5.1.2 Dynamic Augmentation with a Baseline PID Controller

To achieve the PAVs reference dynamics, the helicopter model was augmented with a PID-based controller as shown in Figure 5.1.

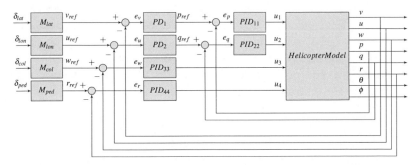

Figure 5.1: Baseline controller to track the TRC specifications of a PAV.

A model following approach was implemented to achieve the response types selected as reference dynamics for each axis (Model reference dynamics [M])[Perfect et al., 2013]. These reference dynamics are the same as those considered in Chapter 3 and reported in

equation 5.2 for convenience. These are Translational Rate Command
(TRC) response types for lateral, longitudinal and vertical translational axes and Rate Command Attitude Hold (RCAH) response type
for the yaw axis.

$$M_{lon} = \frac{u_{ref}}{\delta_{lon}} = \frac{1.05}{1.25s + 1}$$

$$M_{lat} = \frac{v_{ref}}{\delta_{lat}} = \frac{1.05}{1.25s + 1}$$

$$M_{ped} = \frac{r_{ref}}{\delta_{ped}} = \frac{0.03e^{-0.008}}{0.25s + 1}$$

$$M_{col} = \frac{w_{ref}}{\delta_{col}} = \frac{e^{-0.20}}{5s + 1}$$

5.2

As can be seen in Figure 5.1, a multi loop controller architecture
is considered for lateral and longitudinal axes. In the outer loop two
Proportional-Derivative (PD) controllers provide roll and pitch rate
references (p_{ref}, q_{ref}), which are functions of the errors on the lateral
and the longitudinal velocities (e_v, e_u), respectively. The inner loop
tracks the angular rate references and stabilizes the dynamics. A
single loop controller architecture is considered on vertical and yaw
axis. PID_{33} and PID_{44} define the inputs of the helicopter in terms of
errors on the vertical velocity and the yaw rate (e_w, e_r), respectively.

The tuning of the $PIDs$ parameters was done by minimizing
the difference between the augmented helicopter dynamics and the
reference responses (Eq. 5.2), over the frequency range of interest.
Furthermore, the proportional gain of PID_{44} was manually incre-
mented to obtain a better decoupling between vertical velocity and
yaw rate.

The considered control architecture was able to achieve adequate
model tracking within the frequency range of interest, as shown in
Figures 5.2a-5.2d. The result is confirmed in the time domain by
considering step command responses (Figures 5.3a-5.3d). Here it can
be seen that the augmented helicopter behaves approximately like a
first order system, with rising time very close to the PAV reference
response and negligible overshoot. Nevertheless, the responses v to
δ_{lat} and u to δ_{lon} of the augmented system present a delay due to the
non-minimum phase of the helicopter dynamics.

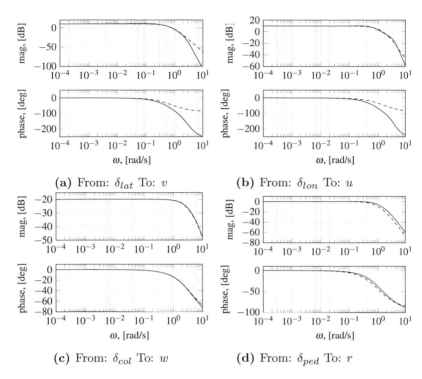

(a) From: δ_{lat} To: v (b) From: δ_{lon} To: u

(c) From: δ_{col} To: w (d) From: δ_{ped} To: r

Figure 5.2: Bode plot comparison. Dashed: Reference Model, Solid: Augmented Helicopter.

5.2   Uncertainties

The identified helicopter model obtained in Chapter 2 is characterized by uncertainties in the identified parameters (ρ). Therefore, each identified parameter can be written in the form:

$$\rho_i = \bar{\rho}_i + \Delta\rho_i \qquad \boxed{5.3}$$

where $\bar{\rho}_i$ represents the nominal value and $\Delta\rho_i$ is the uncertainty associated with the i-th identified parameter, with $i = 1,2,...28$. The expected standard deviations of $\Delta\rho$ are defined by the Cramer-Rao

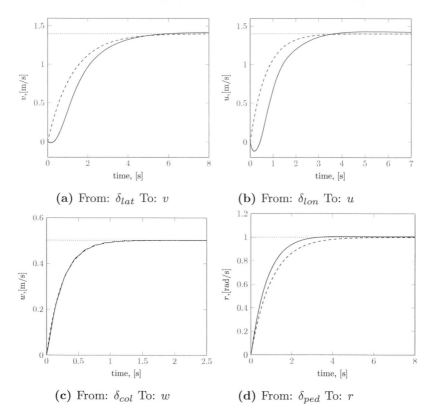

(a) From: δ_{lat} To: v

(b) From: δ_{lon} To: u

(c) From: δ_{col} To: w

(d) From: δ_{ped} To: r

Figure 5.3: Step responses comparison. Dashed: Reference Model, Solid: Augmented Helicopter.

bounds (CR) computed in the state-space model identification. Specifically, $\Delta\rho_i \in [-3CR_i, +3CR_i]$ with 99% of probability [Tischler and Remple, 2012].

Therefore, the uncertain helicopter model dynamics can be written as:

$$\begin{aligned}
\dot{x} &= A_{he}(\bar{\rho} + \Delta\rho) + B_{he}(\bar{\rho} + \Delta\rho)u \\
y &= C_{he}x
\end{aligned}$$

5.4

where the vector of uncertainties $\Delta\rho$ affects both matrices A_{he} and B_{he}. By augmenting the helicopter model with the architecture shown in Figure 5.1, all uncertainties are moved into the state transition

matrix (A_{aug}) of the augmented system. The proof for this is given
in Appendix E. The resulting augmented system can be written in
the form:

$$\dot{x}_{aug} = A_{aug}(\bar{p} + \Delta\rho)x_{aug} + B_{aug}u_{aug}$$
$$y = C_{aug}x_{aug}$$

5.5

where $x_{aug} \in \mathbb{R}^{31}$, $u_{aug} = [v_{cmd}, u_{cmd}, w_{cmd}, r_{cmd}] \in \mathbb{R}^4$ and $y \in \mathbb{R}^8$,
$A_{aug} \in \mathbb{R}^{31x31}$, $B_{aug} \in \mathbb{R}^{31x4}$, $C_{aug} \in \mathbb{R}^{8x31}$. In the following, state
and input vectors of the augmented system will be referred to as x
and u, respectively, to reduce the use of subscripts.

The system in Eq. 5.5 has to be rewritten in a convenient form
in order to apply the adaptive control technique considered in this
chapter. First, the nominal dynamics has to be separated from the
uncertainties:

$$\dot{x} = (A_{aug}(\bar{p}) + \Delta A(\bar{p} + \Delta\rho))x + B_{aug}u$$
$$y = C_{aug}x$$

5.6

where ΔA is unknown and accounts for the uncertainty $\Delta\rho$. Then,
the vector ΔAx is rewritten as a sum of two vectors: the component
of ΔAx along the span of B_{aug} (matched uncertainties) and the
component of ΔAx along the orthogonal complement of B_{aug}, B_{um}
(unmatched uncertainties). Here, $B_{um} \in \mathbb{R}^{31x27}$ is a constant matrix
such that $B_{aug}^T B_{um} = 0$ and $\begin{bmatrix} B_{aug} & B_{um} \end{bmatrix}$ has full rank. The two
components can be expressed as:

$$p_{B_{aug}} = B_{aug}B_{aug}^+\Delta Ax,$$
$$p_{B_{um}} = B_{um}B_{um}^+\Delta Ax.$$

5.7

where the superscript $^+$ indicates the Moore-Penrose pseudoinverse.
As proven in Appendix E, the component of ΔA along the $span(B_{aug})$
is null. Thus, the augmented helicopter dynamics can be written as:

$$\dot{x} = A_{aug}x + B_{aug}u + p_{B_{um}}$$
$$y = C_{aug}x$$

5.8

When the unmatched uncertainties $p_{B_{um}}$ are included in the augmented system, the baseline controller in Figure 5.1 fails to track the
reference model. This is shown in Figures 5.4a-5.4d with the step
responses of $[u, v, w, r]$ with respect to the inputs $[\delta_{lat}, \delta_{lon}, \delta_{col}, \delta_{ped}]$,

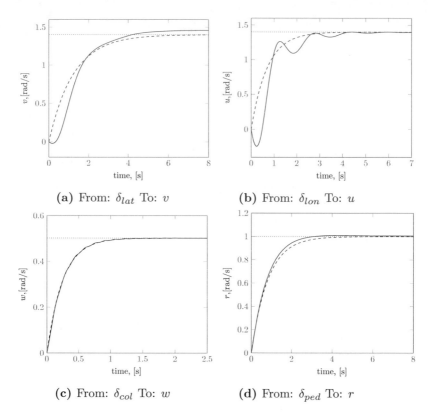

Figure 5.4: Step responses comparison. Solid: Uncertain Augmented Helicopter, Dashed: Reference Model.

respectively. Here, the augmented system was simulated considering one random variation of the uncertain parameters within the boundaries defined by the CR bounds. It is clear how the uncertainties affect the augmented system by making a comparison with respect to the nominal responses shown in Figures 5.3a-5.3d. In particular v due to δ_{lat} presents a long overshoot and u due to δ_{lon} oscillates before reaching the steady-state. Conversely, the responses of w to δ_{col} and r to δ_{ped} are not particularly affected due to the limited amount of uncertainty involved in the vertical and the yaw axes.

These results are reflected in the frequency domain with the
Bode diagrams (Figure 5.5) of the outputs [u, v, w, r] due to the
inputs [δ_{lat}, δ_{lon}, δ_{col}, δ_{ped}], respectively. In this case 50 random
parameter variations are considered. Again, it can be noticed how
the uncertainties mainly affect the lateral and the longitudinal axes
(Figures 5.5a-5.5b), degrading the nominal response.

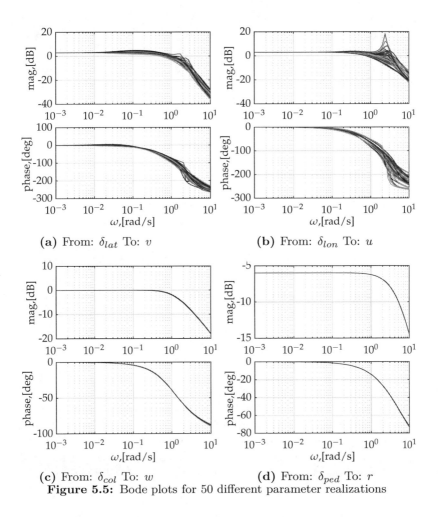

(a) From: δ_{lat} To: v

(b) From: δ_{lon} To: u

(c) From: δ_{col} To: w

(d) From: δ_{ped} To: r

Figure 5.5: Bode plots for 50 different parameter realizations

5.3 $\mathcal{L}_1$-controller theory and implementation

This section gives an overview of $\mathcal{L}_1$-control theory considered here to overcome the detrimental effects of the model parametric uncertainties on the augmented system. In particular, the contents of [Hovakimyan and Cao, 2010] are adapted to the control problem considered here.

The implemented adaptive control architecture is shown in Figure 5.6. Here, the uncertain plant represents the augmented helicopter with uncertainties, the state predictor block computes an estimation of the plant state $\hat{x}$, and the adaptation law block provides an estimation of the uncertainties $\hat{\sigma}$. The control law block creates a control signal based on the reference r and on the uncertainty estimation $\hat{\sigma}$. This control signal has to compensate for the uncertainties effects on the system dynamics. Each block will be now described in details.

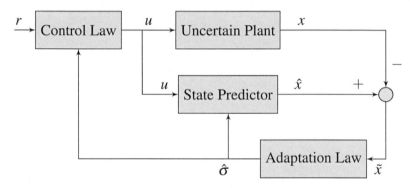

Figure 5.6: Adaptive control architecture.

5.3.1 Uncertain Plant

In literature the adaptive controller is commonly placed around the uncertain element, which corresponds in this case to the helicopter model. However, the compensation for unmatched uncertainties requires the adaptive controller to be applied to a stable systems [Hovakimyan and Cao, 2010]. Since the open loop helicopter model is unstable, it was chosen to apply the adaptive controller to the stable augmented helicopter model.

5.3.2   State Predictor

The *State Predictor* block produces an estimation of the augmented
helicopter state $\hat{x}$, based on the control input of the augmented
helicopter u and the estimation of the unmatched uncertainty $\hat{\sigma}$. The
prediction of the augmented helicopter state is calculated with the
following equation:

$$\dot{\hat{x}}(t) \;=\; A_{aug}\hat{x}(t) + B_{aug}u(t) + B_{um}\hat{\sigma}(t) \qquad \boxed{5.9}$$

where $\hat{\sigma}(t) \in \mathbb{R}^{n-m}$ is the estimation of the unmatched uncertainties
computed in the *Adaptation Law* block. Note that the equations of
the state predictor (Eq. 5.9) and that of the uncertain plant (Eq. 5.8)
have the same form. However, the state predictor dynamics includes
the estimation of the uncertainties.

5.3.3   Adaptation Law

The *Adaptation Law* block provides an estimation of the unmatched uncertainties based on the difference between the state of the augmented
helicopter x and its prediction $\hat{x}$. The estimation of unmatched uncertainties $\hat{\sigma}$ is computed based to the theory presented in [Hovakimyan
and Cao, 2010] as follows:

$$
\begin{aligned}
\hat{\sigma}(t) &\overset{\Delta}{=} \hat{\sigma}_1(t)\|x(t)\|_{L_\infty} + \hat{\sigma}_2(t), \\
\dot{\hat{\sigma}}_1(t) &= -\Gamma(\tilde{x}^T(t)PB_{um})^T\|x(t)\|_{L_\infty}, \\
\dot{\hat{\sigma}}_2(t) &= -\Gamma(\tilde{x}^T(t)PB_{um})^T.
\end{aligned}
\qquad \boxed{5.10}
$$

where $\tilde{x} \overset{\Delta}{=} \hat{x} - x$ is the error between the predicted state and the
actual state of the augmented system. $P \in \mathbb{R}^{nxn}$ is a symmetric,
positive definite matrix that solves the algebraic Lyapunov equation
$A_{aug}^T P + P A_{aug} = -Q$, with $Q \in \mathbb{R}^{nxn}$ definite positive. The choice
of P is related to the proof of stability of the considered control
architecture [Hovakimyan and Cao, 2010]. The design parameter
$\Gamma \in \mathbb{R}^+$ is the adaptive gain. The tuning of Γ is critic for the
implementation of the adaptive controller and will be discussed later
in details.

5.3.4   Control Law

The *Control Law* block produces a control signal that compensates for the effect of unmatched uncertainties using their estimation $\hat{\sigma}$. The estimated effect of unmatched uncertainties on the augmented helicopter outputs can be calculated as $H_{um}(s)\hat{\sigma}$, where

$$H_{um}(s) \overset{\Delta}{=} C_{aug}(s\mathbb{I}_n - A_{aug})B_{um} \in \mathbb{C}^{4x27} \qquad \boxed{5.11}$$

is the transfer function from the unmatched uncertainties to the outputs.

The input that cancels the unmatched uncertainties effect is then calculated as $H_m^{-1}(s)H_{um}(s)\hat{\sigma}(s)$, where $H_m(s)$ is the transfer function that describes the nominal behavior of the augmented helicopter:

$$H_m(s) \overset{\Delta}{=} C_{aug}(s\mathbb{I}_n - A_{aug})B_{aug} \in \mathbb{C}^{4x4} \qquad \boxed{5.12}$$

Finally, the control signal is obtained by adding the computed term for unmatched uncertainties compensation to the reference input $r(s)$ and by applying a low pass filter $C(s)$ as:

$$u(s) = -C(s)(H_m^{-1}(s)H_{um}(s)\hat{\sigma}(s) - r(s)) \qquad \boxed{5.13}$$

In Eq. 5.13, $r(s)$ is the Laplace transform of the pilot input and $C(s) \in \mathbb{C}^{4x4}$ is a proper and stable low-pass filter with unit steady state gain. The choice of the $C(s)$ will also be discussed later in details. To implement the control law defined in Eq. 5.13, $H_{um}(s)$ (Eq. 5.11) must be stable and $H_m(s)$ (Eq. 5.12) must be invertible and have a stable inverse. Here, the stability of $H_{um}(s)$ was ensured by applying the adaptive controller to the stable augmented helicopter. To guarantee an $H_m(s)$ invertible and with stable inverse, a specific outputs selection was necessary. This was done by firstly dividing the outputs of the helicopter model into 4 groups:

- u, q, θ for the longitudinal axis;

- v, p, ϕ for the lateral axis;

- w for the vertical axis;

- r for the yaw axis.

This division was based on the strong dependence among the variables
in each group, due to the under-actuation of the helicopter. Therefore,
selecting one output per control axis was expected to improve the
tracking in all outputs. Based on this considerations, the outputs w,
r, θ, ϕ, were finally selected. These outputs were able to ensure an
existing and stable inverse of $H_m(s)$.

The choice of the adaptation gain Γ and the design of the low-pass
filter $C(s)$ represented the final step. In literature no procedures are
proposed to design Γ and $C(s)$ in multi-input multi-output systems
with unmatched uncertainties [Xargay et al., 2010]. It is well known
that large values of Γ can make the uncertainties estimation fast
but could result in high frequency components, which might lead
to instability. Conversely, low values of Γ produce a slower, yet less
noisy estimation. Generally, choosing a proper $C(s)$ could reduce the
destabilizing effects caused by high values of Γ. Therefore, selecting
a proper filter $C(s)$ can allow for larger values of Γ to be selected.
Based on these considerations, the approach adopted here consisted
of three steps. First, the bandwidth of $C(s)$ was fixed to ensure high
frequencies helicopter dynamics to be maintained. These dynamics
are necessary to compensate for model uncertainties. Second, Γ
was tuned to a high value that allowed for fast estimations while
introducing destabilizing effects. Third, Γ was progressively reduced
until a stable design was achieved. The final parameters obtained
with this approach are listed in Table 5.1.

Parameter	Value
Γ	10^5
$C(s)$	$\frac{1}{(1+\frac{s}{50})^4}$

Table 5.1: Parameter selection

5.4   Monte-Carlo Simulation and Results

The performance of the proposed control architecture was validated via Monte-Carlo simulations. First, a doublet signal (Figure 5.7) was applied to each input of the augmented helicopter without uncertainties. Then, the uncertainties were added and responses to the doublet inputs were evaluated with and without adaptive control. Results were obtained by using 500 random realizations of the identified parametric vector ρ. In particular, as proposed by Tischler [Tischler et al., 2008], for all realizations each parametric perturbation $\Delta\rho_i$ was randomly selected as $\pm 3CR_i$.

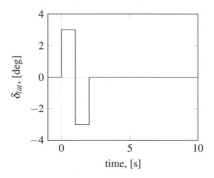

Figure 5.7: Doublet

An example of the obtained results is given in Figures 5.8a-5.8h, where the augmented system outputs are shown due to a doublet on the lateral axis δ_{lat} .

Figures 5.8a-5.8b show the responses of the roll angle ϕ, which is one of the selected outputs directly compensated for uncertainties by the adaptive control. As can be seen, the mean of the responses adequately tracks the nominal response, both with the adaptive controller on and off. However, the use of the adaptive mechanism reduces the standard deviation of the signal.

Figures 5.8c-5.8d show the pitch angle θ responses. It can be noticed that θ is on a coupled secondary axis and is one of the outputs where the adaptation is directly applied. When the adaptive controller is off, the mean response is oscillatory and the standard deviation

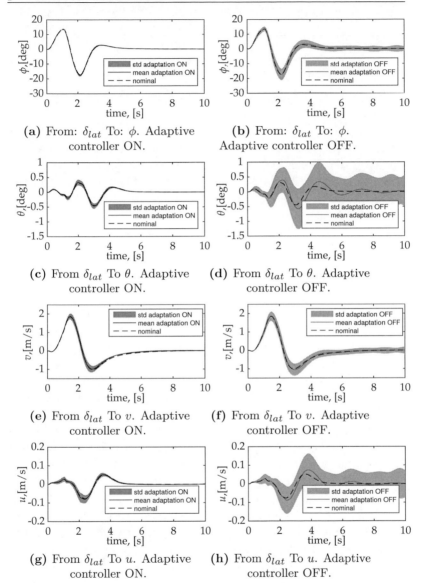

Figure 5.8: Adaptive ON/OFF comparison.

reaches 1 degree. With the adaptation on, both oscillations and standard deviation are significantly reduced.

Figures 5.8e-5.8f show the translational lateral velocity v responses. It can be noticed that although the adaptive controller is not directly applied to v, the adaptation effect is as good as for the compensated outputs.

Finally, Figures 5.8g-5.8h show the translational longitudinal velocity u responses. The output u is not on the primary axis and is not directly compensated by the adaptive controller. However, the adaptation achieves a very good result, as a better tracking of the mean and a much lower standard deviation are obtained than in the case with no adaptation.

Similar results were obtained on the other control axes.

5.5   Conclusions

In this chapter the design of an $\mathcal{L}_1$-adaptive control was proposed as a possible solution for compensating model parametric uncertainties, which degrade the nominal performances of the augmented helicopter model. To verify the validity of the $\mathcal{L}_1$-adaptive control approach, the adaptation was applied to a simple PID-based controller. Furthermore, only translational rate commands (TRC) defined for PAVs were considered as reference dynamics. The PID-based controller was implemented to augment the helicopter model in order to resemble the reference TRC response types. Then, the adaptive control was applied to the augmented helicopter to reject the effects of uncertainties and restore the nominal behavior. Finally, a Monte-Carlo study was performed to validate the proposed control architecture against different realizations of uncertain parameters. Results showed that the implemented adaptive controller is able to significantly reduce the effects of unmatched uncertainties on the augmented helicopter performance. In particular, the adaptive controller is able to exploit the helicopter under-actuation and achieve good tracking of the nominal responses even on outputs not directly compensated. Based on these results, further studies might be considered to apply the $\mathcal{L}_1$-based adaptive control to the augmented control systems presented in Chapter 3.

In this way, the robustness of the implemented H∞ and μ-synthesis control systems could be maintained not only in case of turbulences and noise but also against model parametric uncertainties.

CHAPTER 6

Conclusions and Recommendations

I N 2011 the European Commission funded an out of the box study, the myCopter project [Nieuwenhuizen et al., 2011], with the aim of identifying new concepts for air transport that could be used to achieve a Personal Aerial Transport (PAT) system in the second half of the 21st century. Although designing a new vehicle was not among the project's goal, it was considered important to assess vehicle response types and Handling Qualities that Personal Aerial Vehicles (PAVs) should have to be part of a PAT.

This thesis proposed to consider civil light helicopters as possible PAVs candidates since they possess many properties a PAV should have (e.g. size, weight, number of seats, vertical take-off and landing capabilities). However, civil helicopters are still today a niche product because of costs and long trainings necessary to obtain a pilot license. Conversely, PAVs should be accessible to the general public as today it is for cars.

The goal of this thesis was to investigate whether it is possible to transform civil light helicopters into PAVs through system identification methods and control techniques. The transformation was envisaged in terms of vehicle dynamics and handling qualities. In this way, it was tested whether civil light helicopters can be transformed into vehicles anyone could fly with little training.

To achieve this goal, the thesis was divided into three main steps. The first step, presented in Chapter 2, focused on the implementation of a Robinson R44 Raven II helicopter model in hover. The hover condition was considered well suited for the goal of the thesis as it represents one of the most difficult to perform, particularly for inexperienced pilots. The result of the identification was the development of a linear state-space helicopter model containing rotor-body coupling dynamics important for control designs.

The second step described in Chapter 3 consisted of augmenting the helicopter dynamics to achieve response types and handling qualities defined for PAVs. An optimal H∞ and a robust μ-synthesis techniques were implemented for this goal. Although both controllers were able to resemble PAVs dynamics, even in presence of turbulence and noise, some discrepancies were found between the augmented control systems and the PAV. Furthermore, none of the two control techniques was able to ensure required performance against model parametric uncertainties.

Therefore, a third step was defined to assess the magnitude of discrepancy registered between the two nominal augmented systems and the PAV reference model. An experiment was designed consisting of performing piloted closed-loop control tasks in the MPI CyberMotion Simulator with inexperienced pilots (Chapter 4). Four dynamics were considered in the experiment: the identified helicopter model, the PAV reference model and the two augmented control systems. Each participant was asked to perform two maneuvers (hover and lateral reposition) with one of the selected dynamics. Results were evaluated in terms of objective and subjective workload and performance. In this way, it was possible to compare the two augmented systems with respect to each other, to the original identified helicopter model and to the reference PAV dynamics.

The experiment showed that ordinary inexperienced pilots are not able to control helicopter dynamics. On the contrary, everyone can fly a PAV model and perform maneuvers with good performance. An important result was achieved by the H∞ augmented control system as no significant difference was found with respect to the PAV reference model in terms of objective performance and workload. The μ-synthesis control system performed significantly worse than the H∞

augmented system and than the PAV reference model but only in the vertical axis of the hover MTE.

Overall, both augmented control systems succeeded in resembling PAVs handling qualities and response types in piloted closed-loop control tasks. This result showed that it is possible to transform helicopter dynamics into PAVs ones. Therefore, the approach proposed in this thesis represents a valid alternative to the common practice of implementing new vehicles that can achieve specific requirements like those defined for PAVs.

In Chapter 5 the design of an $\mathcal{L}_1$-adaptive control was proposed to compensate the performance degradation due to the parametric uncertainties of the identified helicopter model. To verify the validity of the $\mathcal{L}_1$-adaptive control approach, the adaptation was applied to a simple PID-based controller. A Monte-Carlo study was performed to validate the proposed control architecture against different realizations of the uncertain parameters. Results showed that the implemented adaptive controller could significantly reduce the detrimental effects of parametric uncertainties on the augmented helicopter performance. Therefore, further studies might be considered to apply the $\mathcal{L}_1$-based adaptive control to the augmented control systems presented in Chapter 3. In this way, the robustness of the implemented H∞ and μ-synthesis control systems could be maintained not only in case of turbulences and noise but also against model parametric uncertainties.

In the next sections a concise overview will be given of the main contributions of this thesis. First, the defined new guidelines for helicopter identification will be considered. Second, the new method implemented for selecting the weighting functions in optimal control designs will be presented. Third, the main findings obtained from the final experiment will be highlighted. A generalization of the main results and recommendations for future works will conclude the thesis.

6.1   Definition of new guidelines for state-space model identification

The first objective of this thesis was the identification of a civil light helicopter model in hover condition. The application of the frequency domain identification method presented by Tischler [Tischler and Remple, 2012] provided good results in the parametric transfer-function case. However, in the state-space case the algorithm used to minimize the MIMO cost function was sensitive to parametric initial values and bounds. As a result of this, different local minima were obtained. The application of iterative pattern search algorithms, usually used to overcome these problems, did not provide satisfactory results as a poor responses fit was obtained (high cost function values). Therefore, a solution was proposed consisting of selecting a larger number of frequency points (20 instead of 50) and a partial coherence function ($\hat{\gamma}^2$) larger than 0.3, instead of 0.5. Although this approach was not based on theoretical analyses, it was considered that an increase of the frequency content could be achieved by including points with lower coherence. During the minimization, the frequency points were weighted with the associated coherence values. This procedure resulted more effective at finding solutions not sensitive to initial parametric values and bounds.

However, low coherence functions are generally associated with process noise, nonlinearities, lack of input excitation or lack of rotorcraft response. Therefore, the original guidelines (20 frequency points and partial coherence $\hat{\gamma}^2 > 0.5$) were again applied to fit the final state-space model. This time though, local minima were avoided by selecting initial parametric values and bounds based on the solutions provided by the new guidelines approach.

The new guidelines approach limited the sensitivity of the minimization algorithm to initial parametric values and bounds. The obtained model presented good predictive capabilities and was able to capture high frequency rotor-body coupling dynamics. The model was also validated and positively assessed by an expert helicopter pilot while performing maneuvers in hover and low speed flight regime in the MPI CyberMotion Simulator.

6.2   Weighting function selection in optimal control design

The second objective of this thesis was to augment the identified civil helicopter model to achieve response types and handling qualities defined for PAVs. To achieve this goal, two robust control approaches were considered, the H∞ and μ-synthesis techniques. In these control approaches specific weighting functions have to be selected. Usually, selecting weighting functions necessary to achieve an optimal solution is a crucial step, in which different aspects have to be taken into account. Here, the weighting functions zeros and poles were fixed to satisfy requirements concerning frequency range of interest, control effort and bandwidths linked to the actuators performance.

The gains were instead optimized to ensure a satisfactory tracking of the reference responses. The optimization was not trivial as classical procedures did not provide controllers able to achieve satisfactory performance results. For example, a stronger control action obtained by limiting the related weighting functions gains was not effective because large control gains on longitudinal and lateral axes generated instability. Furthermore, ensuring good tracking of both translational and rotational PAV reference responses was not straightforward as in a helicopter these dynamics are strictly coupled.

To overcome these issues and satisfy all requirements, a multi-objective optimization problem with competing objectives was defined and solved through the application of a genetic algorithm (Chapter 3). The obtained optimal solution provided weighting functions able to guarantee good reference tracking and to resemble PAVs handling qualities.

The optimization problem was defined for the H∞ control problem and was also applied to the μ-synthesis control design, as it was computationally harder to find ad-hoc optimal weights for this control technique.

6.3   Assessment of augmented control systems in the MPI CyberMotion Simulator

The last objective of this thesis was to assess the implemented control systems in terms of objective and subjective performance and workload. Therefore, an experiment was implemented in which participants without any prior flight experience were asked to perform piloted closed loop control tasks in the MPI CyberMotion Simulator. Four dynamics were tested: the identified R44 helicopter model, the PAV reference model and the two augmented systems ($H\infty$ and μ-synthesis). Every participant performed two types of maneuvers, adapted versions of the hover and the lateral reposition MTEs defined in the ADS-33E document [Baskett, 2000]. An actual helicopter pilot was also invited to take part in the experiment. The pilot was assigned to the experimental condition involving the control of the helicopter model. This was done to allow for comparisons with the other participants.

The results of the experiment provided important information about the four selected dynamics. In the helicopter model condition, six out of seven participants were not able to stabilize the assigned dynamics and perform the two selected maneuvers. This showed that ordinary inexperienced pilots cannot control a helicopter dynamics. However, one inexperienced subject was able to control the helicopter dynamics and to perform both maneuvers. This result was attributed to the previous subject experience with control tasks involving helicopter-like dynamics in simulation. Furthermore, the subject was able to quickly learn how to control the dynamics by taking advantage of the adopted experimental approach. In this approach, the possibility to crash and immediately start a new trial accelerated the learning process with respect to the classical training adopted for helicopters. Furthermore, suggestions given by the experimenter during and after each trial allowed subjects to learn from previous mistakes and to quickly improve their performance. Since this approach was beneficial to only one participant in the helicopter dynamics condition, further studies and experiments should be conducted to investigate which factors are crucial for allowing inexperienced pilots to learn how to control a helicopter. Furthermore, it could be studied whether few hours of flight tests in a motion simulator might bring benefits before flying on an actual helicopter.

The second important result was that all participant who were assigned to the experimental conditions involving the PAV model or one of the two augmented systems were able to control the dynamics and perform both selected maneuvers. Furthermore, important insights were provided into the features of the augmented systems and the magnitude of discrepancy with respect to the PAV reference model. The measures used for objective performance and workload showed no significant difference between the H∞ augmented system and the PAV reference model in both MTEs. The μ-synthesis augmented system showed a decrease in performance in the hover MTE with respect to the PAV model. However, a significant difference was obtained only in the vertical axis. Therefore, it was concluded that both augmented systems succeeded in resembling PAVs response types and handling qualities in piloted closed-loop control tasks. This result positively answered the main question asked in this thesis as to whether civil helicopters can be transformed into PAVs.

A third important result of the experiment was that the actual helicopter pilot achieved higher performance than the medians of all other groups of participants. This was expected, given the pilot's previous experience with actual helicopters and flight simulators. Nevertheless, this result showed that ordinary PAVs pilots would need longer trainings than the one given in this experiment to achieve performance comparable to those of experienced helicopter pilots.

A final important result was obtained from the objective workload measure. In the collected data no statistical difference was found between the two augmented systems and the PAV reference model. However, a larger amount of control activity was necessary to perform the maneuvers for both inexperienced subjects and actual helicopter pilot in the helicopter dynamics condition.

The NASA TLX questionnaire did not provide any additional finding to evaluate hypotheses and to answer questions considered in this thesis. This was due to the experimental design, conceived as a "between subjects" one. In this way, participants had the chance to practice and perform maneuvers with only one of the four possible dynamics. Therefore, they rated the assigned dynamics without considering any reference for comparisons. As such, subjects focused on different aspects like the difficulty at controlling simultaneously four control devices while flying in a 3 dimensional space. This led to some discrepancies in the subjective ratings with respect to the objective workload measures.

6.4   Generalization of the results

This thesis proposed a new method for implementing Personal Aerial Vehicles that could be adopted in a Personal Aerial Transport system. Starting from an existing vehicle, the civil light helicopter, it was demonstrated that identification methods and control techniques can be successfully applied to achieve response types and handling qualities defined for PAVs. The main questions that motivated this research were fully answered. Although the results shown here are bound by the scope of the thesis, some of them can be generalized.

An example is the choice of the hover and the low speed flight maneuvers. These flight conditions were considered well suited for the goal of the thesis as they are very difficult to perform, particularly for inexperienced pilots. Therefore, all results and findings presented in the thesis could be generalized and extended to other flight conditions (e.g. high speed task maneuvers) generally considered less demanding for inexperienced pilots.

Conversely, other findings cannot be directly generalized. In this thesis only robust control techniques were implemented as considered well suited for MIMO linear systems affected by uncertainties, external disturbances and sensor noise. Although these techniques were assessed in terms of robustness and performance and compared to each other, a generalized evaluation with respect to other control methods (e.g. PID, inverse control dynamics etc.) cannot be directly considered without performing specific experiments.

The control augmented systems were evaluated with an experiment in the MPI CyberMotion simulator by considering a virtual environment with calm air and good external visual conditions. This choice was made to compare the augmented systems with respect to the original identified helicopter model and to the PAV reference model without including the effects of external factors. However, a real flight scenario can present wind gusts, vibrations or degraded visual conditions. These factors could influence performance and workload. Therefore, the obtained results cannot be directly extended to all possible real flight conditions without including them into an experiment.

Finally, a training of 20 minutes was selected for the last experiment. This choice was based on previous results, which had proven such amount of training to be sufficient for allowing inexperienced pilots to control the PAV dynamic model. Therefore, the same amount of training was assigned to all experimental conditions for comparison. The final experiment showed that only one inexperienced pilot was able to control the helicopter dynamics at the end of the training. This result does not allow for generalizations on the amount of training necessary to allow inexperienced pilots to learn how to control a helicopter. In fact, all other participants did not achieve satisfactory results until the end of the experiment.

6.5 Recommendations

The results presented in this thesis validated the proposed method of creating PAVs through the augmentation of civil light helicopters. Furthermore, some of these results opened important questions that could be investigated in the future.

The two control methods considered in this thesis allowed an identified helicopter dynamics to resemble response types and HQs defined for PAVs. However, the results presented in Chapter 3 highlighted the incapability of both augmented systems of maintaining nominal performance against model parametric uncertainties. This issue was faced in Chapter 5 with the implementation of a $\mathcal{L}_1$-based adaptive control. To verify the effectiveness of this approach, the adaptation was applied to a simple PID-based control. Furthermore, only the translational rate commands defined for PAVs were considered as reference dynamics. Although this method was proven to be well suited for this problem, further studies should be considered to apply the

$\mathcal{L}_1$-based adaptive control to the H∞ and the μ-synthesis augmented systems implemented in this thesis.

Another interesting study would be to investigate how model parametric uncertainties can affect performance and workload in piloted closed loop control tasks. To do this, an experiment could be conducted in which inexperienced pilots would perform MTEs for different parametric values of the uncertain augmented dynamics. This could allow the definition of HQs levels for inexperienced pilots as those defined in the ADS-33E-PRF for trained pilots.

The last experiment showed that one inexperienced pilot in the helicopter group was able to control the helicopter dynamics after 20 minutes of training. Conversely, all other participants of the same group were not able to achieve the same result until the end of the experiment. Based on these results, a study could be considered to investigate which factors are crucial to help inexperienced pilots learn how to control a helicopter. This could allow inexperienced pilots to receive specific trainings in which these factors would be adapted to accelerate the learning process.

Bibliography

"Agard lecture series 178, rotorcraft system identification," Tech. rep., AGARD Advisory Group for Aerospace Research & Development, 7 Rue Ancelle, 92200 Neuilly Sur Seine, France, 1995.

"Unity 3d," 2015.
http://unity3d.com/

Alan Simpson, V. B. and Stoker, J., *UAS Safety in Non-segregated Airspace*, INTECH Open Access Publisher, 2009, doi:10.5772/6492.
http://www.intechopen.com/books/aerial_vehicles/uas_safety_in_non-segregated_airspace

Baskett, B. J., "Aeronautical Design Standard Performance Specification Handling Qualities Requirements for Military Rotorcraft," Tech. Rep. ADS-33E-PRF, United States Army Aviation and Missile Command, Aviation Engineering Directorate, Redstone Arsenal (AL), 2000.

Bendat, J. S. and Piersol, A. G., *Random Data: Analysis and Measurement Procedures, Fourth Edition*, John Wiley & Sons, 2010.

Bibel, J. E. and Malyevac, S. D., "Guidelines for the selection of weighting functions for h-infinity control," Tech. Rep. AD-A251 781, DTIC Document, 1992.

Bichlmeier, M., Holzapfel, F., Xargay, E., and Hovakimyan, N., "L1 adaptive augmentation of a helicopter baseline controller," *AIAA Guidance, Navigation and Control Conference, Boston, Massachusetts*, 2013.

Boyd, S. P. and Barratt, C. H., *Linear controller design: limits of performance*, Prentice Hall Englewood Cliffs, NJ, 1991.

Bülthoff, H. H., Nieuwenhuizen, F. M., and Padfield, G., "mycopter: Enabling technologies for personal air transport systems," *Proceedings of the RAeS Rotorcraft Conference: The Future Rotorcraft-Enabling Capability Through the Application of Technology*, pp. 1–11, The Royal Aeronautical Soc. London, 2011.

Chen, R. T. N. and Hindson, W. S., "Influence of higher-order dynamics on helicopter flight control system band-width," *Journal of Guidance, Control, and Dynamics*, vol. 9, no. 2, pp. 190–197, 1986.

Dorobantu, A., Murch, A., Mettler, B., and Balas, G., "System identification for small, low-cost, fixed-wing unmanned aircraft," *Journal of Aircraft*, vol. 50, no. 4, pp. 1117–1130, 2013.

Fletcher, J. W., "Identification of uh-60 stability derivative models in hover from flight test data," *Journal of the American Helicopter Society*, vol. 40, no. 1, pp. 32–46, 1995a.

Fletcher, J. W., "A model structure for identification of linear models of the uh-60 helicopter in hover and forward flight," Technical report 95-a-008, NASA Technical Memorandum 110362 USAATCOM, 1995b.

Fu, K. H. and Kaletka, J., "Frequency-domain identification of bo 105 derivative models with rotor degrees of freedom," *Journal of the American Helicopter Society*, vol. 38, no. 1, pp. 73–83, Jan. 1993.

Geluardi, S., Nieuwenhuizen, F. M., Pollini, L., and Bülthoff, H. H., "Data collection for developing a dynamic model of a light helicopter," *Proceedings of the 39th European Rotorcraft Forum 2013, Moscow*, 2013.

Geluardi, S., Nieuwenhuizen, F. M., Pollini, L., and Bülthoff, H. H., "Frequency domain system identification of a light helicopter in hover," *Proceedings of the American Helicopter Society 70th Annual Forum, Montreal, Canada*, May 20-22 2014.

Gregory, I. M., Cao, C., Xargay, E., Hovakimyan, N., and Zou, X., "L1 adaptive control design for nasa airstar flight test vehicle," *AIAA guidance, navigation, and control conference, Pasadena, California*, vol. 5738, 2009.

Greiser, S. and von Gruenhagen, W., "Analysis of model uncertainties using inverse simulation," *Proceedings of the AHS 69th Annual Forum, Phoenix, Arizona*, May 21-23 2013.

Greiser, S. and Lantzsch, R., "Equivalent modelling and suppression of air resonance for the act/fhs in flight," *Proceedings of the 39th European Rotorcraft Forum 2013, Moscow*, 2013.

Guerreiro, B. J., Silvestre, C., Cunha, R., Cao, C., and Hovakimyan, N., "L1 adaptive control for autonomous rotorcraft," *Proceedings of the American Control Conference, 2009, St. Louis, Missouri*, pp. 3250–3255, IEEE, 2009.

Ham, J. A., Gardner, C. K., and Tischler, M. B., "Flight testing and frequency domain analysis for rotorcraft handling qualities," *Journal of the American Helicopter Society*, vol. 40, no. 2, pp. 28–38, Apr. 1995.

Hamel, P. G. and Kaletka, J., "Advances in rotorcraft system identification," *Progress in Aerospace Sciences*, vol. 33, pp. 259–284, Mar.-Apr. 1997.

Hart, S. G. and Staveland, L. E., "Development of nasa-tlx (task load index): Results of empirical and theoretical research," *Advances in psychology*, vol. 52, pp. 139–183, 1988.

Heffley, R. K., "A compilation and analysis of helicopter handling quality data," Tech. Rep. NASA-CR-314, National Aeronautics and Space Administration, Scientific and Technical Information Branch, 1979.

Hovakimyan, N. and Cao, C., *L1 adaptive control theory: guaranteed robustness with fast adaptation*, vol. 21, Siam, 2010.

Ivler, C. and Tischler, M., "Case studies of system identification modeling for flight control design," *Journal of the American Helicopter Society*, vol. 58, no. 1, pp. 1–16, Jan. 2013.

Jategaonkar, R. V., Fischenberg, D., and Gruenhagen, W., "Aerodynamic modeling and system identification from flight data-recent applications at dlr," *Journal of Aircraft*, vol. 41, no. 4, pp. 681–691, 2004.

Kaletka, J., Kurscheid, H., and Butter, U., "Fhs, the new research helicopter: Ready for service," *Aerospace science and technology*, vol. 9, no. 5, pp. 456–467, 2005.

Klein, V. and Morelli, E. A., *Aircraft system identification: Theory and Practice*, AIAA American Institute of Aeronautics and Astronautics Reston, VA, USA, 2006.

Lavretsky, E. and Wise, K., *Robust and adaptive control: with aerospace applications*, Springer Science & Business Media, 2012.

Moller, P. S., "Airborne personalized travel using powered lift aircraft," *AIAA and SAE World Aviation Conference*, vol. 107, pp. 1572–1578, 1998.

Moore, M., "Nasa puffin electric tailsitter vtol concept," *10th AIAA Aviation Technology, Integration, and Operations (ATIO) Conference*, pp. 13–15, AIAA Reston, VA, 2010.

Nieuwenhuizen, F. M. and Bülthoff, H. H., "The mpi cybermotion simulator: A novel research platform to investigate human control behavior," *JCSE*, vol. 7, no. 2, pp. 122–131, 2013.

Nieuwenhuizen, F. M., Jump, M., Perfect, P., White, M., Padfield, G., Floreano, D., Schill, F., Zufferey, J., Fua, P., Bouabdallah, S., Siegwart, R., Meyer, S., Schippl, J., Decker, M., Gursky, B., Höfinger, M., and Bülthoff, H. H., "mycopter: Enabling technologies for personal aerial transportation systems," *Proceedings of the 3rd International HELI World Conference, Frankfurt/Main*, 2011.

Perfect, P., Jump, M., and White, M. D., "Investigation of personal aerial vehicle handling qualities requirements for harsh environmental conditions," *Proceedings of the American Helicopter Society 70th Annual Forum, Montreal, Canada*, May 20-22 2014.

Perfect, P., Jump, M., and White, M. D., "Handling qualities requirements for future personal aerial vehicles," *Journal of Guidance, Control, and Dynamics*, pp. 1–13, 2015a.

Perfect, P., Jump, M., and White, M. D., "Methods to assess the handling qualities requirements for personal aerial vehicles," *Journal of Guidance, Control, and Dynamics*, 2015b.

Perfect, P., White, M. D., and Jump, M., "Towards handling qualities requirements for future personal aerial vehicles," *Proceedings of the American Helicopter Society 69th Annual Forum, Phoenix, Arizona*, May 21-23 2013.

Postlethwaite, I., Prempain, E., Turner, E. T. M. C., Ellis, K., and Gubbels, A. W., "Design and flight testing of various h-infinity controllers for the bell 205 helicopter," *Control Engineering Practice*, vol. 13, no. 3, pp. 383–398, 2005.

Quiding, C., Ivler, C. M., and Tischler, M. B., "Genhel s-76c model correlation using flight test identified models," *Proceedings of the American Helicopter Society 64th Annual Forum, Montreal, Canada*, 29 April - 1 May 2008.

Robinson, "R44 ii pilot's operating handbook," Tech. rep., Robinson Helicopter Company, 1992.

Sahai, R., Cicolani, L., Tischler, M., Blanken, C., Sullivan, C., Wei, M., Ng, Y.-S., and Pierce, L., "Flight-time identification of helicopter-slung load frequency response characteristics using cifer," *Proceedings of the 24th Atmospheric Flight Mechanics Conference*, August 1999.

Seher-Weiss, S. and von Gruenhagen, W., "Development of ec 135 turbulence models via system identification," *Aerospace Science and Technology*, vol. 23, no. 1, pp. 43–52, 2012.

Skogestad, S. and Postlethwaite, I., *Multivariable feedback control: analysis and design*, vol. 2, Wiley New York, 2007.

Theodore, C. R., Malpica, C. A., Blanken, C. L., Tischler, M. B., Lawrence, B., Lindsey, J. E., and Berger, T., "Effect of control system augmentation on handling qualities and task performance in good and degraded visual environments," *Proceedings of the American Helicopter Society 70th Annual Forum, Montreal, Canada*, May 20-22 2014.

Theodore, C. R., Tischler, M. B., and Colbourne, J. D., "Rapid frequency-domain modeling methods for unmanned aerial vehicle flight control applications," *Journal of Aircraft*, vol. 41, no. 4, pp. 735–743, 2004.

Tischler, M. B., "System identification requirements for high-bandwidth rotorcraft flight control system design," *Proceeding of the American Control Conference*, 1991.

Tischler, M. B., "System identification methods for aircraft flight control development and validation," Tech. Rep. NASA-TM-110369, Aeroflightdynamics Directorate, U.S. Army ATCOM, Ames Research Center, Moffett Field, CA 94035-10008, Oct. 1995.

Tischler, M. B. and Cauffman, M. G., "Frequency response method for rotorcraft system identification: Flight applications to bo 105 coupled rotor/fuselage dynamics," *Journal of the American Helicopter Society*, vol. 37, no. 3, pp. 3–17, Jul. 1992.

Tischler, M. B., Ivler, C. M., Mansur, M. H., Cheung, K. K., Berger, T., and Berrios, M., "Handling-qualities optimization and trade-offs in rotorcraft flight control design," *AHS Specialists Meeting on Rotorcraft Handling-Qualities, Liverpool, UK*, pp. 4–6, 2008.

Tischler, M. B., Lee, J. A., and Colbourne, J. D., "Comparison of flight control system design methods using the conduit design tool," *Journal of Guidance, Control, and Dynamics*, vol. 25, no. 3, pp. 482–493, 2002.

Tischler, M. B. and Remple, R. K., *Aircraft and Rotorcraft System Identification Engineering Methods with Flight Test Examples*, AIAA Education Series, 2006.

Tischler, M. B. and Remple, R. K., *Aircraft and Rotorcraft System Identification Engineering Methods with Flight Test Examples*, AIAA Education Series, 2012.

Truman, T. and de Graaff, A., "Out of the box. ideas about the future of air transport part 2," Tech. rep., EUROPEAN COMMISSION Directorate General for Research, November 2007.

Walker, D. J. and Postlethwaite, I., "Advanced helicopter flight control using two-degree-of-freedom h-infinity optimization," *Journal of Guidance, Control, and Dynamics*, vol. 19, no. 2, pp. 461–468, 1996.

Walker, D. J., Turner, M. C., and Gubbels, A., "Practical aspects of implementing h-infinity controllers on a fbw research helicopter," *RTO AVT Symposium, Braunschweig, Germany*, May 8-11 2000.

Wax, H., "Ready for takeoff," *Women in Engineering Magazine, IEEE*, vol. 4, no. 1, pp. 14–17, 2010.

Williams, J. N., Ham, J. A., and Tischler, M. B., "Flight test manual, rotorcraft frequency domain flight testing," Tech. rep., US Army Aviation Technical Test Center, Edwards AFB, CA, AQTD Project, No. 93-14, 1995.

Wiskemann, C., Drop, F., Pool, D., van Paassen, M., Mulder, M., and Bülthoff, H., "Subjective and objective metrics for the evaluation of motion cueing fidelity for a roll-lateral reposition maneuver," *Proceedings of the American Helicopter Society 70th Annual Forum, Montreal, Canada*, May 20-22 2014.

Xargay, E., Hovakimyan, N., and Cao, C., "L1 adaptive controller for multi-input multi-output systems in the presence of nonlinear unmatched uncertainties," *American Control Conference, Marriott Waterfront, Baltimore, Maryland*, pp. 874–879, 2010.

Zhou, K. and Doyle, J. C., *Essentials of robust control*, vol. 104, Prentice hall Upper Saddle River, NJ, 1999.

Zivan, L. and Tischler, M. B., "Development of a full flight envelope helicopter simulation using system identification," *Journal of the American Helicopter Society*, vol. 55, no. 2, p. 22003, Apr. 2010.

Appendices

APPENDIX A

Procedure for flight test data collection

A technical procedure was developed for the data collection presented in Chapter 2. The main steps of this procedure were described in a technical report that was used to instruct the technicians on the setup to mount and the helicopter pilot on the maneuvers to perform during the flight tests. The document starts with a brief description of the main steps for the collection of data and the relative estimated time necessary to implement them. At the end of the document some instructions are given for the flight-test engineer on the software to be used to collect the data. The descriptions of the maneuvers to perform are reported from the book [Tischler and Remple, 2012].

**Technical procedure for the collection of flight test data
with a R44 helicopter**

Stefano Geluardi

Procedure main steps	TIME	Paragraph
Setup preparation. Measurement of IMU and GPS positions. Measurement of optical sensors positions. Measurement of plates positions and slopes.	1.5 h	1.
Piloted control mapping. Consider different positions. Leave the control in each position for few second. Save the mapping data on the pc.	30 min	2.
Compute the CoG position using the knowledge of the position and weight of equipment and people onboard during the flight	15 min	3.
Pilot training phase: theoretical description of the maneuvers. Training with display on the ground of doublets and frequency sweeps.	30 min	4.
Flight test	1 h	5.

1. Mounting of the setup

1) Place the optical sensors on the four pilot controls. Place the reference plates in front of the sensors.
2) Place the IMU and measure its position with respect to the front of the helicopter and with respect to the skid.
3) Place the 2 GPS antennas on the left skid and measure their position with respect to the IMU.
4) Place PC, display and battery on the rear seat of the helicopter and measure their positions and weights.
5) Connect the cable without label to the port COM2 of the PC.
6) Connect the cable to the device used for the second GPS antennas (COM2 Flex-Pack).
7) Connect the 2 CANBus to the T connection. Connect the Port CAN1 of the PC.
8) Open the file CONFIG2 on the desktop.
9) Insert the position of the 2 GPS antennas with respect to the IMU.
10)   Type in: log rawimusa onnew.
11)   Type in: canconfig can1 enable 1M 200 83FB insgps.
12)   The setup is ready for the data collection.

2. Control piloted mapping procedure

In this procedure the angular displacements of the pilot controls are measured. Furthermore, the linear distance from the sensors to the reference plates is to be measured in order to compute the mapping from angles to distance.

For each axis:

1) Connect the measuring device.
2) Calibrate the measuring device.
3) Perform the following procedure:

Measurement procedure for the cyclic

1) Start by finding the (0,0) pitch and roll values.
2) Move the stick to the extreme forward position. Then, move the stick towards the extreme back position holding it 3 seconds in each of 10 different intermediate positions.
3) Move the stick to the extreme left position. Then, move the stick towards the extreme right position holding it 3 seconds in each of 10 different intermediate positions.
4) Collect some stick positions close to the center (position (0,0)).
5) Perform some realistic piloted maneuvers for validation.

Measurement procedure for collective and pedals

6) Move the control from one the extreme position to the other position holding it 3 seconds in each of 10 different intermediate positions.

3. Computation of the center of gravity position

The computation of the CoG position should be done after the placement of the setup by knowing the exact position and weight of all the equipment and of the people on board during the flight.

The CoG position should be done by using the software provided by Robinson for the R44 Raven II helicopter. Only the longitudinal and the lateral positions can be computed through the software. The vertical CoG position has to be identified during the identification process.

4. Pilot training phase on the ground

The pilot training phase is performed on the ground and in flight.

First, a theoretical description of the specific maneuvers is given to the pilot to make him aware of the kind of movements he has to perform for each control axis.

Then, a first the training is conducted on the ground to instruct the pilot to perform the selected maneuvers with correct input timing and magnitude.

A display is used to show to the pilot the performed maneuvers in real time.

Two types of maneuvers are to be performed: doubles and frequency sweeps.

Doublets

A doublet on one control axis can be performed as follows:

the control is moved in one direction until a considered angular rate or linear velocity is achieved. Then, the control input is reversed.

The piloted input has to be roughly symmetric as in the picture.

Frequency sweeps

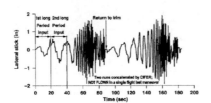

What is important

1. *Sweeps* should start and end in **trim**, with **3 s of trim**
2. After the initial trim period, execute **two complete long-period inputs** T_{max}
3. Maintain a **smooth increasing progression in frequency**, without rushing through the mid-frequencies
4. Adjust the input to maintain the aircraft response transients to be **roughly symmetric about the trim flight condition**

5. **Nonswept controls (step and pulse-type control inputs)** are applied to "bound" the off-axis responses to be roughly symmetric with respect to the reference flight condition

6. **Providing timing indicators** to the pilot to assist in the frequency-sweep tests

What is NOT important

1. **Constant input** amplitude is NOT important
2. **Exact sinusoidal** input shape is NOT important
3. **Exact frequency progression** is NOT important
4. **Exact repeatability** is not important and indeed is NOT desired
5. **Increased input amplitudes** at higher frequencies are NOT needed
6. **High-frequency inputs** are NOT needed

Displacement: NO MORE than 10%-20% of the control inputs range

Typically 0.5-1 inches ➔ ±5 m/s ; ±15 deg ; ±10 kn

Frequency: NO MORE than 2 Hz (Higher frequencies could cause structural damages).

Example of lateral stick piloted sweep (the rules are the same for all the control axes)

The maneuver starts with 3 seconds of trim.

The "flight-test engineer" should call out "5, 10, 15, 20" to signal the pilot when the lateral stick position should be roughly at maximum right, center, maximum left, and center for the first long-period cycle. The movement should be very smooth.

The flight test continues with "25, 30, 35, 40" for the second long-period cycle.

At this point the frequency should slowly increase. This could be done by calling "44, 48, 52, 56" end so on towards the higher frequencies.

The time length of the entire sweep is around 90 seconds as you can see in the picture.

The maneuver ends with 3 seconds of trim.

5. Flight test

The flight test is performed in the following way:

1) Weather conditions check for flying: temperature, wind, density, altitude.
2) Alignment procedure for the INS while performing some maneuvers in ground effect.
3) Start data recording.
4) Doublets maneuvers with cyclic control (longitudinal and lateral axis)
5) Frequency sweeps with cyclic control (longitudinal; and lateral axis)
6) Checking on the ground of performed maneuvers(consistency analysis, frequency of interest, amplitude, timing, coherence function input-input, coherence function input-output of primary responses)
7) Performing at least 3 good frequency sweeps and 2 good doublets for each control axis (longitudinal cyclic, lateral cyclic, collective, pedals).

Appendix A

Data consistency and reconstruction

1)   Kinematic consistency (angular and translational)

2)   Correction instrumentation system characteristics. Mechanical and filter characteristics transformed into time delay (e^{-ts})

3)   Detection of faulty data.

4)   Use of the Extended Kalman Filter (EKF) for kinematic consistency checking.

Software to use

1)   Use parse_helidata.m for saving messages (modify the recording-name required)

3)   Use var_def.m for defining all variables names of interest (add the optical sensor variables)

4)   Interpolation of data and data drop out elimination

5)   Concatenation of data maneuvers

6)   detrend.m

7)   Use CoG_correction.m

8)   Check the data with the EKF and imu_calc.m

9)   plot_input_output.m

10)   bode_coherence.m

Measurement setup for data collection

THE collection of data presented in Chapter 2 was obtained through the measurement setup reported in Figure B.1.   Four optical

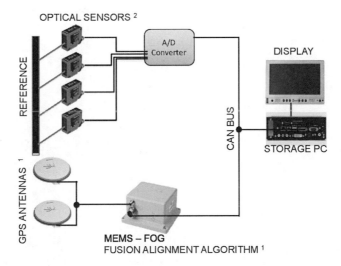

Figure B.1: Schematic overview of the measurement setup used for the flight tests.

[1]http://www.novatel.com/products/span-gnss-inertial-systems/span-combined-systems/span-cpt

[2]http://www.wenglor.com

sensors were attached to the pilot controls to measure their displacement with respect to plastic surfaces used as references. An inertial measurement unit (IMU) was connected to two Global Positioning Systems (GPS) antennas to provide linear accelerations, angular velocities and vehicle inertial position. The IMU comprised Fiber Optic Gyros (FOG) and Micro Electrical Mechanical System (MEMS) accelerometers. A storage PC was used to save all measured data with a sample rate of 100 Hz.

The characteristics of the sensors used in the measurement setup are listed in Table B.1 in terms of resolutions and ranges.

Table B.1: Instrumentation properties

Sensor	Resolution	Range
Accelerometers[1]	$0.005\ m/s^2$	$\pm10\ g$
Gyro Output[1]	$0.01\ deg/s$	$\pm375\ deg/s$
Optical sensors CP24MHT80[2]	$<20\ \mu m$	$40\text{-}160\ mm$
Optical sensors CP35MHT80[2]	$<50\ \mu m$	$50\text{-}350\ mm$

[1] novatel.com/assets/Documents/Papers/SPAN-CPT.pdf
[2] http://www.wenglor.com

APPENDIX C

State space identification frequency responses

IN this appendix the primary frequency responses are shown, obtained from the state-space model identification presented in Chapter 2. For each response the relative magnitude, phase and coherence are presented. Figures C.1 and C.2 show the primary responses of the lateral axis due to the lateral cyclic input δ_{lat}. Figures C.3 and C.4 show the primary responses of the longitudinal axis due to the longitudinal cyclic input δ_{lon}. Figure C.5 shows the primary responses of the vertical axis due to the collective input δ_{col}. Finally, Figure C.6 shows the primary response of the yaw axis due to the pedals input δ_{ped}.

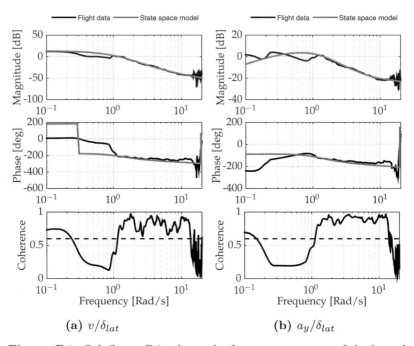

(a) v/δ_{lat} **(b)** a_y/δ_{lat}

Figure C.1: Sub-figure C.1a shows the frequency response of the lateral translational velocity to the lateral cyclic input (v/δ_{lat}). Sub-figure C.1b shows the the frequency response of the lateral acceleration to the lateral cyclic input (a_y/δ_{lat}).

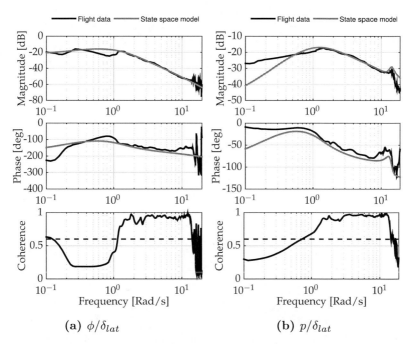

(a) ϕ/δ_{lat} **(b)** p/δ_{lat}

Figure C.2: Sub-figure C.2a shows the frequency response of the roll angle to the lateral cyclic input (ϕ/δ_{lat}). Sub-figure C.2b shows the the frequency response of the roll-rate to the lateral cyclic input (p/δ_{lat}).

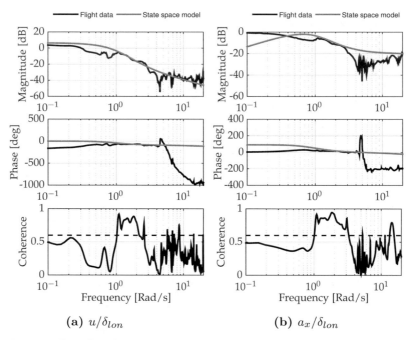

(a) u/δ_{lon} **(b)** a_x/δ_{lon}

Figure C.3: Sub-figure C.3a shows the frequency response of the translational longitudinal velocity to the longitudinal cyclic input (u/δ_{lon}).
Sub-figure C.3b shows the the frequency response of the translational
longitudinal acceleration to the longitudinal cyclic input (a_x/δ_{lon}).

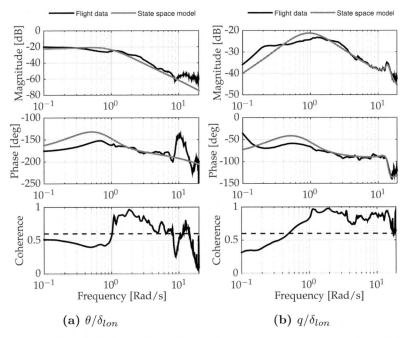

(a) θ/δ_{lon} **(b)** q/δ_{lon}

Figure C.4: Sub-figure C.4a shows the frequency response of the pitch angle to the longitudinal cyclic input (θ/δ_{lon}). Sub-figure C.4b shows the the frequency response of the pitch-rate to the longitudinal cyclic input (q/δ_{lon}).

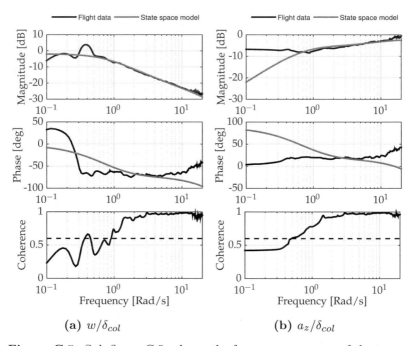

(a) w/δ_{col} **(b)** a_z/δ_{col}

Figure C.5: Sub-figure C.5a shows the frequency response of the translational vertical velocity to the collective input (w/δ_{col}). Sub-figure C.5b shows the the frequency response of the vertical acceleration to the collective input (a_z/δ_{col}).

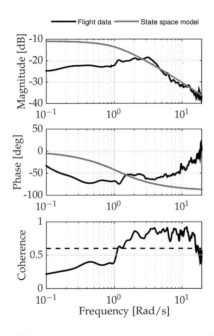

Figure C.6: Frequency response yaw-rate to pedals input (r/δ_{ped}).

APPENDIX D

Experiment briefing

B EFORE starting the experiment described in Chapter 4, each participant received a briefing provided in this appendix. An extensive
oral briefing was also given about the objectives of the experiment, the
tasks to perform and the experimental procedures. Participants were
invited to ask questions during the entire experiment and to provide
suggestions at the end of it. The briefing included a questionnaire
to rate the participants state before and after the experiment and
check for motion sickness symptoms. At the end of the briefing, a
sketch was provided of the maneuvers to perform taken from the
ADS-33E-PRF document [Baskett, 2000].

Max Planck Institute for Biological Cybernetics
Human Perception, Cognition and Action
Spemannstr. 38, D-72076 Tübingen

MAX-PLANCK-GESELLSCHAFT   BIOLOGISCHE KYBERNETIK

Informed Consent (stipend holders or external only)

I, _____, state that I am _____ years of age and that I agree to participate in a research study being conducted by Stefano Geluardi of the Max Planck Institute for Biological Cybernetics.

I acknowledge that _____ has informed me that my participation in this study is voluntary, that I may withdraw my participation at any time without penalty and without having to give a reason. All data that I contribute will remain confidential and will be coded so that the anonymity will be protected in any research papers and presentations that result from this work. I understand that in this experiment I will be presented with vehicle simulations inside the CyberMotion Simulator. I will receive a complete explanation of the study's goals at the end of my participation. I understand that the study involves no serious risk. For participating in the study I will receive a compensation of 8 Euro/hour at the end of the study. If I decide to not complete the study, or if the study is interrupted for technical reasons, I will be paid for my participation time until the interruption of the study.

_____   _____

(Signature of Participant)   (Date)

_____   _____

(Signature of Researcher)   (Date)

If you would like to receive a summary of the results at the conclusion of the study, please write your email address here: _____

THANK YOU FOR PARTICIPATING!

Contact: Stefano Geluardi | Max Planck Institute for Biological Cybernetics | Human Perception, Cognition and Action | Spemannstr. 38, D-72076 Tübingen |Tel: 07071 601 607| Email: stefano.geluardi@tuebingen.mpg.de

Max Planck Institute for Biological Cybernetics
Human Perception, Cognition and Action
Spemannstr. 38, D-72076 Tübingen

MAX-PLANCK-GESELLSCHAFT

BIOLOGISCHE KYBERNETIK

Baseline Questionnaire

Subject ID:	Condition:	Date:	Time:

1. Please indicate in what level you experience the following symptoms (encircle):

	low			high
General discomfort	1	2	3	4
Fatigue	1	2	3	4
Headache	1	2	3	4
Eyestrain	1	2	3	4
Difficulty focussing (visually)	1	2	3	4
Increased salivation	1	2	3	4
Sweating	1	2	3	4
Nausea	1	2	3	4
Difficulty concentrating	1	2	3	4
Fullness of the head	1	2	3	4
Blurred vision	1	2	3	4
Dizzy[1] (eyes open)	1	2	3	4
Dizzy[1] (eyes closed)	1	2	3	4
Vertigo[2] (i.e., spinning)	1	2	3	4
Stomach awareness	1	2	3	4

[1] Dizziness is a feeling of light-headedness or equilibrium imbalance
[2] Vertigo is the feeling that either you or the surroundings are spinning.

2. How would you rate your current state?

| 1 | 2 | 3 | 4 | 5 | 6 | 7 | 8 | 9 |

(e.g.)   (e.g.)
tired   energetic
demotivated   motivated
distracted   concentrated
weak   fit
ill   healthy

General comments:

....

....

....

Contact: Stefano Geluardi | Max Planck Institute for Biological Cybernetics | Human Perception, Cognition and Action | Spemannstr. 38, D-72076 Tübingen |Tel: 07071 601 607| Email: stefano.geluardi@tuebingen.mpg.de

Max Planck Institute for Biological Cybernetics
Human Perception, Cognition and Action
Spemannstr. 38, D-72076 Tübingen

Evaluation Questionnaire

Subject ID:	Condition:	Date:	Time:

The following questions assess your mental and physical state during the simulator session you just experienced.

1. Please indicate in what level you experienced the following symptoms (encircle):

	low			high
General discomfort	1	2	3	4
Fatigue	1	2	3	4
Headache	1	2	3	4
Eyestrain	1	2	3	4
Difficulty focussing (visually)	1	2	3	4
Increased salivation	1	2	3	4
Sweating	1	2	3	4
Nausea	1	2	3	4
Difficulty concentrating	1	2	3	4
Fullness of the head	1	2	3	4
Blurred vision	1	2	3	4
Dizzy[1] (eyes open)	1	2	3	4
Dizzy[1] (eyes closed)	1	2	3	4
Vertigo[2] (i.e., spinning)	1	2	3	4
Stomach awareness	1	2	3	4
Burping	1	2	3	4

[1] Dizziness is a feeling of light-headedness or equilibrium imbalance
[2] Vertigo is the feeling that either you or the surroundings are spinning.

General comments:

....

....

....

Please continue on the following page

Contact: Stefano Geluardi | Max Planck Institute for Biological Cybernetics | Human Perception, Cognition and Action | Spemannstr. 38, D-72076 Tübingen |Tel: 07071 601 607| Email: stefano.geluardi@tuebingen.mpg.de

Max Planck Institute for Biological Cybernetics
Human Perception, Cognition and Action
Spemannstr. 38, D-72076 Tübingen

Hover Maneuver

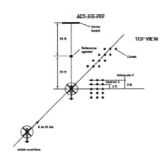

Lateral Maneuver

Figure 26. Suggested course for sidestep and vertical remask maneuvers

Contact: Stefano Geluardi | Max Planck Institute for Biological Cybernetics | Human Perception, Cognition and Action | Spemannstr. 38, D-72076 Tübingen | Tel: 07071 601 607 | Email: stefano.geluardi@tuebingen.mpg.de

Mathematical proofs for adaptive control design

IN this appendix the state space model of the system in Figure 5.1 is considered. The helicopter state-space model is given in Eq. 5.1 where the feedback outputs $[u,\ v,\ w,\ p,\ q,\ r]$ are selected. The PID-based controller model can be expressed in state-space form as follows:

$$\begin{array}{rcl} \dot{x}_c & = & A_c x_c + B_{c1} y + B_{c2} r \\ u & = & C_c x_c + D_{c1} y + D_{c2} r \end{array}$$

E.1

where $x_c \in \mathbb{R}^{n_c}$ is the controller state vector, u the control input from 5.1, y is the feedback output from 5.1, and $r \in \mathbb{R}^{n_r}$ represents the reference vector. The real matrices $(A_c,\ B_{c1},\ B_{c2},\ C_c,\ D_{c1},\ D_{c2})$ are of appropriate dimensions and describe the controller dynamics.

The state-space model for the closed loop system of helicopter and PID-based controller is:

$$\begin{array}{rcl} \dot{x}_{cl} = \begin{bmatrix} x_{he} \\ \dot{x}_c \end{bmatrix} & = & A_{cl} x_{cl} + B_{cl} r \\ y & = & C_{cl} x_{cl} + D_{cl} r \end{array}$$

E.2

where matrices $(A_{cl},\ B_{cl},\ C_{cl},\ D_{cl})$ are:

$$A_{cl}(\rho) = \begin{bmatrix} A_{he}(\rho) + B_{he}(\rho)D_{c1}C_{he} & B_p(\rho)C_{he} \\ B_{c1}C_{he} & A_c \end{bmatrix}$$

$$B_{cl}(\rho) = \begin{bmatrix} B_{he}(\rho)D_{c2} \\ B_{c2} \end{bmatrix}$$

$$C_{cl} = C_{he}$$

$$D_{cl} = 0$$

E.3

As shown in Figure 5.1 the reference model [M] is in cascade with the system in E.3. Since the reference model is strictly proper, its state space form is:

$$\dot{x}_r = A_r x_r + B_r u_r$$
$$r = C_r x_r$$

E.4

where $x_r \in \mathbb{R}^{n_r}$ is the state, $u_r \in \mathbb{R}^{m_r}$ is the external command vector and, r is the reference vector from E.1. The real matrices $(A_r,\ B_r,\ C_r)$ are of appropriate dimensions and describe the reference dynamics. Finally, the cascade of the reference model [M] and the helicopter augmented with the $PID-based$ controller can be expressed in state space form as follows:

$$\dot{x} = \begin{bmatrix} \dot{x}_{cl} \\ \dot{x}_r \end{bmatrix} = Ax + Bu_r$$
$$y = Cx + Du_r$$

E.5

where matrices $(A,\ B,\ C,\ D)$ are:

$$A(\rho) = \begin{bmatrix} A_{cl}(\rho) & B_{cl}(\rho)C_r \\ 0 & A_r \end{bmatrix} \quad B = \begin{bmatrix} 0 \\ B_r \end{bmatrix}$$

$$C = C_{cl} \qquad\qquad D = 0$$

E.6

From E.6 it can be observed that:

- Matrix B only depends on the reference dynamics and thus is not affected by uncertainties;

- All uncertainties lie in the upper rows of A and matrix B in null in the corresponding rows. Thus, all uncertainties are unmatched since they are not in the span of B.

Acknowledgments

This PhD thesis was possible only thanks to the contribution of many important People in my life.

Firstly, I am grateful to God, or how I like to call him "Turiddu". This PhD presented many challenges that I faced with serenity only because I knew He was always there next to me.

I wish to express my sincere thanks to my supervisors, Heinrich Bülthoff and Lorenzo Pollini for giving me the opportunity to conduct my PhD research as a collaboration between the Max Planck Institute for Biological Cybernetics and the University of Pisa. They supported my research with sincere and valuable guidance and encouragement and they provided me with facilities that cannot be found elsewhere.

I am also immensely grateful to Frank and Joost for their supervision and their unique capability of reasoning in a clear and straightforward way. From them, I learned many competencies that are important to become a scientist with a critical but always constructive point of view.

I would also like to thank my friends and colleagues at the MPI, in particular Carlo, Ale, Burak, Menja, Christiane, Eva, Lewis, Janina, Giacomo and Giulia, for sharing with me this amazing experience and for all the fun we have had in the last four years.

A special big thank goes to Mario and Frank for the many interesting discussions in our office, not always related with our work but most of the time characterized by a highly scientific and philosophical content.

Finally, I would like to express my gratitude to my parents, my brothers and my grandparents, for their faith in me and their countless prayers.

And, of course, I would like to express my special appreciation and thanks to my beloved Simona, for supporting every single moment of my life and my research, even from far away.